RELIGION AND THE RISE OF CAPITALISM

# RELIGION

## AND THE

# RISE OF CAPITALISM

## A HISTORICAL STUDY

### (HOLLAND MEMORIAL LECTURES, 1922)

### By R. H. TAWNEY
READER IN ECONOMIC HISTORY, UNIVERSITY OF LONDON;
SOMETIME FELLOW OF BALLIOL COLLEGE, OXFORD

GLOUCESTER, MASS.
PETER SMITH
1962

"Whatever the world thinks, he who hath not much meditated upon God, the human mind, and the *summum bonum,* may possibly make a thriving earthworm, but will most indubitably make a sorry patriot and a sorry statesman."

BISHOP BERKELEY, *Siris,* 350.

# INTRODUCTION

THE object of this book is to trace some strands in the development of religious thought on social and economic questions in the period which saw the transition from medieval to modern theories of social organization. It does not carry the subject beyond the beginning of the eighteenth century, and it makes no pretense of dealing with the history either of economic theory or of economic practice, except in so far as theory and practice were related to changes in religious opinion. In reality, however, the connection between them was intimate and vital. The revolutions, at once religious, political and social, which herald the transition from the medieval to the modern world, were hardly less decisive for the economic character of the new civilization than for its ecclesiastical organization and religious doctrines. The economic categories of modern society have their roots in the economic expansion and social convulsions which accompanied the age of the Renaissance and the Reformation.

The history of religious thought on questions of social ethics is a topic which has been treated in England by the late Dr. Cunningham, by Sir William Ashley, whose essay on *The Canonist Doctrine* first interested me in the subject, by Mr. G. G. Coulton, Mr. H. G. Wood, and Mr. G. O'Brien. But it is no reflection on their work to say that the most important contributions of recent years have come from continental students, in particular Troeltsch, Choisy, Sombart, Brentano, Levy and, above all, Max Weber, whose celebrated essay on *Die Protestantische Ethik und der Geist des Kapitalismus* gave a new turn to the discussion. No one can work, on however humble a scale, in the same field, without being conscious of the heavy obligation under which these scholars have laid him. While I have not always been able to accept their conclusions, I am glad to have this opportunity of expressing my indebtedness to them. I regret that Mr. Coulton's *The Mediæval Village* appeared too late for me to make use of its abundant stores of learning and insight.

It only remains for me to thank the friends whose assistance has enabled me to make this book somewhat less imperfect than it would otherwise have been. Mr. J. L. Hammond, Dr. E. Power, and Mr. A. P. Wadsworth have been kind enough to read, and to improve, the manuscript. Professor J. E. Neale, in addition to reading the proofs, has helped me most generously throughout with advice and criticism. I am deeply indebted both to Miss Bulkley, who has undertaken the thankless task of correcting the proofs and making an index, and to the London School of Economics and the Laura Spelman Rockefeller Memorial Fund for enabling me to make use of her services. My obligation to the help given by my wife is beyond acknowledgment.

R. H. TAWNEY.

# PREFACE TO 1937 EDITION

SINCE the appearance of this book ten years ago, the literature on its subject has considerably increased. The learned
work of Troeltsch, the best introduction to the historical
study of religious thought on social issues, can now be read
in an English translation, as can also the articles of Weber
on *The Protestant Ethic and the Spirit of Capitalism.* The
omission from my book of any reference to post-Reforma-
tion Catholic opinion was a serious defect, which subsequent
writers have done something to repair. The development
of economic thought in mediaeval Italy; the social forces at
work in the Germany of Luther, and his attitude to them;
the economic doctrines of Calvin; the teaching of the Jesuits
on usury and allied topics; English social policy during the
Interregnum; the religious and social outlook of the French
*bourgeoisie* of the same period; the attitude of Quakers,
Wesleyans, and other bodies of English Nonconformists to
the changing economic world which confronted them in the
eighteenth century, have all had books devoted to them. In
the somewhat lengthy list of articles on these and kindred
subjects, those by the late Professor Sée, M. Halbwachs,
and Mr. Parsons, and an article by Mr. Gordon Walker
which has just appeared in *The Economic History Review,*
specially deserve attention.[1]

It will be seen, therefore, that the problems treated in the
following pages, if they continue to perplex, have not ceased
to arouse interest. What conclusions, if any, emerge from
the discussion?

The most significant are truisms. When this book first
appeared, it was possible for a friendly reviewer, writing in

a serious journal, to deprecate in all gravity the employment of the term "Capitalism" in an historical work, as a political catch-word, betraying a sinister intention on the part of the misguided author.  An innocent solecism of the kind would not, it is probable, occur so readily to-day.  Obviously, the word "Capitalism," like "Feudalism" and "Mercantilism," is open to misuse.  Obviously, the time has now come when it is more important to determine the different species of Capitalism, and the successive phases of its growth, than to continue to labour the existence of the genus.  But, after more than half a century of work on the subject by scholars of half a dozen different nationalities and of every variety of political opinion, to deny that the phenomenon exists; or to suggest that, if it does exist, it is unique among human institutions, in having, like Melchizedek, existed from eternity; or to imply that, if it has a history, propriety forbids that history to be disinterred, is to run wilfully in blinkers. Verbal controversies are profitless; if an author discovers a more suitable term, by all means let him use it.  He is unlikely, however, to make much of the history of Europe during the last three centuries, if, in addition to eschewing the word, he ignores the fact.

The more general realization of the role of Capitalism in history has been accompanied by a second change, which, if equally commonplace, has also, perhaps, its significance. "Trade is one thing, religion is another": once advanced as an audacious novelty, the doctrine that religion and economic interests form two separate and co-ordinate kingdoms, of which neither, without presumption, can encroach on the other, was commonly accepted by the England of the nineteenth century with an unquestioning assurance at which its earliest exponents would have felt some embarrassment. An historian is concerned less to appraise the validity of an idea than to understand its development.  The effects for good or evil of that convenient demarcation, and the forces which, in our own day, have caused the boundary to shift, need not here be discussed.  Whatever its merits, its victory, it is now realized, was long in being won.  The economic theories propounded by Schoolmen; the fulminations by the left wing of the Reformers against usury, landgrabbing, and extortionate prices; the appeal of hard-headed Tudor states-men to traditional religious sanctions; the attempt of Calvin and his followers to establish an economic discipline more rigorous than that which they had overthrown, are bad evi-dence for practice, but good evidence for thought.  All rest on the assumption that the institution of property, the trans-actions of the market-place, the whole fabric of society and the whole range of its activities, stand by no absolute title, but must justify themselves at the bar of religion.  All in-sist that Christianity has no more deadly foe than the *ap-petitus divitiarum infinitus,* the unbridled indulgence of the acquisitive appetite.  Hence the claim that religion should keep its hands off business encountered, when first formu-lated, a great body of antithetic doctrine, embodied not only in literature and teaching, but in custom and law.  It was only gradually, and after a warfare not confined to paper, that it affected the transition from the status of an odious paradox to that of an unquestioned truth.

The tendency of that transition is no longer in dispute. Its causation and stages remain the subject of debate.  The critical period, especially in England, was the two centuries following the Reformation.  It is natural, therefore, that most recent work on the subject of this book should have turned its high lights on that distracted age.  The most striking attempt to formulate a theory of the movement of religious thought on social issues which then took place was made at the beginning of the present century by a German scholar, Max Weber,[2] in two articles published in 1904 and 1905.  Hence it is not less natural that much of that work

should, consciously or unconsciously, have had Weber as its
starting-point.

What exactly was the subject with which he was concerned? That question is obviously the first which should
be asked, though not all his critics ask it. He was preparing
to undertake the comparative study of the social outlook and
influence of different religions, the incomplete results of
which appeared in three volumes in 1920, under the name
of *Gesammelte Aufsätze zur Religionssoziologie.* The articles, *Die protestantische Ethik und der Geist des Kapitalismus,* were a first step towards that larger work, and subsequently, corrected and amplified, formed part of its first
volume. Weber thought that western Christianity as a
whole, and in particular certain varieties of it, which acquired an independent life as a result of the Reformation,
had been more favourable to the progress of Capitalism
than some other great creeds. His articles were an attempt
to test that generalization.

Their scope is explained in an introduction written later
to the *Religionssoziologie.* His object was to examine—the
abstractions fall with a mournful thud on English ears—
"the influence of certain religious ideas on the development
of an economic spirit or the *ethos* of an economic system."
He hoped—*O sancta simplicitas!*—to avoid misunderstand-
ing by underlining somewhat heavily the limitations of his
theme. He formulated no "dogma"; on the contrary, he
emphasized that his articles were to be regarded as merely a
*Vorarbeit,*[3] a preparatory essay. He did not seek "a psychological determination of economic events";[4] on the contrary, he insisted on "the fundamental importance of the
economic factor."[5] He did not profess to offer a complete
interpretation even of the religious attitude discussed in his
articles; on the contrary, he urged the necessity of investi-
gating how that attitude itself "was in turn influenced in its
development and character by the totality of social condi-

tions, especially the economic ones." [6]   So far from desiring
—to quote his own words—"to substitute for a one-sided
'materialistic' an equally one-sided 'spiritual' interpretation
of civilization and history," [7] he expressly repudiated any
intention of the kind.

In view of these disclaimers, it should not be necessary
to point out that Weber made no attempt in the articles in
question to advance a comprehensive theory of the genesis
and growth of Capitalism.   That topic had been much discussed in Germany since Marx opened the debate, and the
first edition of the most massive of recent books on the
subject, Sombart's *Der Moderne Kapitalismus,* had appeared two years before.   The range of Weber's interests,
and the sweep of his intellectual vision, were, no doubt,
unusually wide; but his earliest work had been done on economic history, and he continued to lecture on that subject
till his death in 1920.   If he did not in his articles refer to
the economic consequences of the discovery of America, or
of the great depreciation, or of the rise to financial pre-emi-
nence of the Catholic city of Antwerp, it was not that these
bashful events had at last hit on an historian whose notice
they could elude.   Obviously, they were epoch-making; obviously, they had a profound effect, not only on economic
organization, but on economic thought.   Weber's immediate
problem, however, was a different one.   Montesquieu remarked, with perhaps excessive optimism, that the English
"had progressed furthest of all people in three important
things, piety, commerce and freedom."   The debt of the
third of these admirable attributes to the first had often
been emphasized.   Was it possible, Weber asked, that the
second might also owe something to it?   He answered that
question in the affirmative.   The connecting link was to be
found, he thought, in the influence of the religious movement whose greatest figure had been Calvin.

Since Weber's articles are now available in English, it is

needless to recapitulate the steps in his argument. My own
views upon it, if I may refer to them without undue egotism, were summarized in a note—too lengthy to be read—
to the first edition of the present work, and were later restated more fully in the introduction to the English translation of the articles which appeared in 1930.[8] Weber's
generalizations had been widely discussed by continental
scholars for more than twenty years before this book appeared. The criticisms contained in it, therefore, had no
claim to originality—unless, indeed, to be less anxious to
refute an author than to understand him is in itself to be
original.

The first of them—that "the development of Capitalism
in Holland and England in the sixteenth and seventeenth
centuries was due, not to the fact that they were Protestant
Powers, but to large economic movements, in particular the
Discoveries and the results which flowed from them"—has
since been developed at some length by Mr. Robertson; but
it was not, perhaps, quite just. Weber would have replied,
no doubt, that such a remark, however true, was, as far as
his articles were concerned, an *ignoratio elenchi*. To meet
him fairly, he would have said, one should meet him on his
own ground, which at the moment was that, not of general
economic history, but of religious thought on social issues.
My second comment, already made by Brentano—that more
weight should have been given to the political thought of
the Renaissance—had been anticipated by Weber,[9] and I
regret that I overlooked his observations on that point. His
gravest weaknesses in his own special field, where alone
criticism is relevant, are not those on which most emphasis
has usually been laid. The Calvinist applications of the
doctrine of the "Calling" have, doubtless, their significance;
but the degree of influence which they exercised, and their
affinity or contrast with other versions of the same idea, are
matters of personal judgment, not of precise proof. Both

Weber and his critics have made too much of them, as I did myself. His account of the social theory of Calvinism, however, if it rightly underlined some points needing emphasis, left a good deal unsaid. The *lacunae* in his argument cannot here be discussed, but two of them deserve notice. Though some recent attempts to find parallels to that theory in contemporary Catholic writers have not been very happy, Weber tended to treat it as more unique than it was.[10] More important, he exaggerated its stability and consistency. Taking a good deal of his evidence from a somewhat late phase in the history of the movement, he did not emphasize sufficiently the profound changes through which Calvinism passed in the century following the death of Calvin.

The last point is of some moment. It suggests that the problem discussed by Weber requires to be restated. It is natural, no doubt, that much of the later work on the subject should have taken him for its target, and probably inevitable—such is the nature of controversy—that a theory which he advanced as a hypothesis to explain one range of phenomena, and one alone, should have been clothed for the purpose of criticism with the uncompromising finality of a remorseless dogma. His mine has paid handsome dividends; but, whatever its attractions, that vein, it may be suggested, is now worked out. The important question, after all, is not what Weber wrote about the facts, still less what the *epigoni* who take in his washing have suggested that he wrote, but what the facts were. It is an illusion to suppose that he stands alone in pointing to a connection between the religious movements of the sixteenth and seventeenth centuries and the outburst of economic energy which was remaking society in the Netherlands and England. Other students have reached, independently of him, that not recondite conclusion.[11] How much truth does it contain?

To attempt a reply to that question would expand a preface into a book. The materials for answering it are, how-

ever, abundant. If contemporary opinion on the point is
not easily cited, the difficulty arises, not from lack of evidence to reveal it, but from the *embarras de richesse* which
it offers for quotation. Its tenor is not doubtful. The
truth is that the ascription to different confessions of distinctive economic attitudes was not exceptional in the seventeenth century; among writers who handled such topics it
was almost common form. It occurs repeatedly in works
of religious controversy. It occurs also in books, such as
those of Temple, Petty, and Defoe, and numerous pamphlets, by men whose primary interest was, not religion, but
economic affairs. So far, in fact, from being, as has been
suggested [12] with disarming *naïveté,* the sinister concoction
of a dark modern conspiracy, designed to confound Calvinism and Capitalism, godly Geneva and industrious Manchester, in a common ruin, the existence of a connection between economic Radicalism and religious Radicalism was to
those who saw both at first-hand something not far from
a platitude. Until some reason is produced for rejecting
their testimony, it had better be assumed that they knew
what they were talking about.

How precisely that connection should be conceived is, of
course, a different question. It had, obviously, two sides.
Religion influenced, to a degree which to-day is difficult to
appreciate, men's outlook of society. Economic and social
changes acted powerfully on religion. Weber, as was natural in view of his special interests, emphasized the first
point. He did so with a wealth of knowledge and an intellectual force which deserve admiration, and not least the
admiration of those who, like myself, have ventured to dissent from some of his conclusions. He touched the second
point only *en passant.* There is truth in the criticism of Mr.
Gordon Walker that Weber did not inquire how far the
Reformation was a response to social needs, or investigate

the causes, as well as the consequences, of the religious mentality which he analysed with so much insight.

It is that aspect of the subject which most needs work to-day. In the triple reconstruction, political, ecclesiastical, and economic, through which England passed between the Armada and the Revolution, every ingredient in the caldron worked a subtle change in every other. There was action and reaction. "L'esprit calviniste," and "l'esprit des hommes nouveaux que la révolution économique du temps introduit dans la vie des affaires," [18] if in theory distinct, were in practice intertwined. Puritanism helped to mould the social order, but it was also itself increasingly moulded by it. Of the influence of the economic expansion of the age on English religious thought something is said in the following pages. I hope that their inadequacies may prompt some more competent writer to deal with that subject as its importance deserves.

R. H. TAWNEY

# CONTENTS

# CHAPTER I

## THE MEDIEVAL BACKGROUND

"La miséricorde de Dieu est infinie: elle sauvera même un riche."
ANATOLE FRANCE, *Le Puits de Sainte Claire.*

# CHAPTER I

## THE MEDIEVAL BACKGROUND

"QUE pourrions-nous gagner," once wrote a celebrated economist, "à recueillir des opinions absurdes, des doctrines décriées, et qui méritent de l'être? Il serait à la fois inutile et fastidieux de les exhumer." [1] One who studies the development of social theory can hardly hope to avoid the criticism which is brought against those who disturb the dust in forgotten lumber-rooms. If he seeks an excuse beyond his own curiosity, he may find it, perhaps, in the reflection that the past reveals to the present what the present is capable of seeing, and that the face which to one age is a blank may to another be pregnant with meaning. Writing when economic science was in the first flush of its dogmatic youth, it was natural that Say should dismiss as an unprofitable dilettantism an interest in the speculations of ages unillumined by the radiance of the new Gospel. But to determine the significance of opinion is, perhaps, not altogether so simple a matter as he supposed. Since the brave days when Torrens could say of Political Economy, "Twenty years hence there will scarcely exist a doubt respecting any of its fundamental principles," [2] how many confident certainties have been undermined! How many doctrines once dismissed as the emptiest of superstitions have revealed an unsuspected vitality!

The attempt to judge economic activity and social organization by ethical criteria raises problems which are eternal, and it is possible that a study of the thought of an age when that attempt was made, if with little success, at least with conviction and persistence, may prove, even today, not wholly without instruction.    In the present century, the old issues seem, indeed, to have acquired a new actuality.    The philosophy which would keep economic interests and ethical idealism safely locked up in their separate compartments finds that each of the prisoners is increasingly restive.    On the one hand, it is evident that the whole body of regulations, by which modern societies set limits to the free play of economic self-interest, implies the acceptance, whether deliberate or unconscious, of moral standards, by reference to which certain kinds of economic conduct are pronounced illegitimate.    On the other hand, there are indications that religious thought is no longer content to dismiss the transactions of business and the institutions of society as matters irrelevant to the life of the spirit.

Silently, but unmistakably, the conception of the scope and content of Christian ethics which was generally, though not universally, accepted in the nineteenth century, is undergoing a revision; and in that revision the appeal to the experience of mankind, which is history, has played some part, and will play a larger one.    There have been periods in which a tacit agreement, accepted in practice if not stated in theory, excluded economic activities and social institutions from examination or criticism in the light of religion. A statesman of the early nineteenth century, whose conception of the relations of Church and State appears to have been modeled on those of Mr. Collins and Lady Catherine de Bourgh, is said to have crushed a clerical reformer with the protest, "Things have come to a pretty pass if religion is going to interfere with private life"; and a more recent occupant of his office has explained the catastrophe which must follow, if the Church crosses the Rubicon which divides the outlying provinces of the spirit from the secular capital of public affairs.[3]

Whatever the merit of these aphorisms, it is evident today that the line of division between the spheres of religion and secular business, which they assume as self-evident, is shifting. By common consent the treaty of partition has lapsed and the boundaries are once more in motion. The age of which Froude, no romantic admirer of ecclesiastical pretensions, could write, with perhaps exaggerated severity, that the spokesmen of religion "leave the present world to the men of business and the devil," [4] shows some signs of drawing to a close. Rightly or wrongly, with wisdom or with its opposite, not only in England but on the Continent and in America, not only in one denomination but among Roman Catholics, Anglicans, and Nonconformists, an attempt is being made to restate the practical implications of the social ethics of the Christian faith, in a form sufficiently comprehensive to provide a standard by which to judge the collective actions and institutions of mankind, in the sphere both of international politics and of social organization. It is being made today. It has been made in the past. Whether it will result in any new synthesis, whether in the future at some point pushed farther into the tough world of practical affairs men will say,

> Here nature first begins
> Her farthest verge, and chaos to retire
> As from her outmost works, a broken foe,

will not be known by this generation. What is certain is that, as in the analogous problem of the relations between Church and State, issues which were thought to have been buried by the discretion of centuries have shown in our own day that they were not dead, but sleeping. To examine the forms which they have assumed and the phases through which they have passed, even in the narrow field of a single country and a limited period, is not mere antiquarianism. It is to summon the living, not to invoke a corpse, and to see from a new angle the problems of our own age, by widening the experience brought to their consideration.

In such an examination the sixteenth and seventeenth centuries are obviously a critical period. Dr. Figgis [5] has described the secularization of political theory as the most momentous of the intellectual changes which ushered in the modern world. It was not the less revolutionary because it was only gradually that its full consequences became apparent, so that seeds which were sown before the Reformation yielded their fruit in England only after the Civil War. The political aspects of the transformation are familiar. The theological mould which shaped political theory from the Middle Ages to the seventeenth century is broken; politics becomes a science, ultimately a group of sciences, and theology at best one science among others. Reason takes the place of revelation, and the criterion of political institutions is expediency, not religious authority. Religion, ceasing to be the master-interest of mankind, dwindles into a department of life with boundaries which it is extravagant to overstep.

The ground which it vacates is occupied by a new institution, armed with a novel doctrine. If the Church of the Middle Ages was a kind of State, the State of the Tudors had some of the characteristics of a Church; and it was precisely the impossibility, for all but a handful of sectaries, of conceiving a society which treated religion as a thing privately vital but publicly indifferent, which in England made irreconcilable the quarrel between Puritanism and the monarchy. When the mass had been heated in the furnace of the Civil War, its component parts were ready to be disengaged from each other. By the end of the seventeenth century the secular State, separate from the Churches, which are subordinate to it, has emerged from the theory which had regarded both as dual aspects of a single society. The former pays a shadowy deference to religion; the latter do not meddle with the external fabric of the political and social system, which is the concern of the former. The age of religious struggles virtually ends with the Treaty of Westphalia in 1648. The age of the wars of economic nationalism virtually begins with the war between England and Holland under the Commonwealth and Charles II. The State, first in England, then in France and America, finds its sanction, not in religion, but in nature, in a presumed contract to establish it, in the necessity for mutual protection and the convenience of mutual assistance. It appeals to no supernatural commission, but exists to protect individuals in the enjoyment of those absolute rights which were vested in them by the immutable laws of nature. "The great and chief end of men uniting into commonwealths and putting themselves under government is the preservation of their property." [6]

While the political significance of this development has often been described, the analogous changes in social and economic thought have received less attention. They were, however, momentous, and deserve consideration. The emergence of an objective and passionless economic science took place more slowly than the corresponding movement in the theory of the State, because the issues were less absorbing, and, while one marched in the high lights of the open stage, the other lurked on the back stairs and in the wings. It was not till a century after Machiavelli had emancipated the State from religion, that the doctrine of the self-contained department with laws of its own begins generally to be applied to the world of business relations, and even in the England of the early seventeenth century, to discuss questions of economic organization purely in terms of pecuniary profit and loss still wears an air of not quite reputable cynicism. When the sixteenth century opens, not only political but social theory is saturated with doctrines drawn from the sphere of ethics and religion, and economic phenomena are expressed in terms of personal conduct, as naturally and inevitably as the nineteenth century expressed them in terms of mechanism.

Not the least fundamental of divisions among theories of society is between those which regard the world of human affairs as self-contained, and those which appeal to a supernatural criterion.  Modern social theory, like modern political theory, develops only when society is given a naturalistic instead of a religious explanation, and a capital fact which presides at the birth of both is a change in the conception held of the nature and functions of a Church.  The crucial period is the sixteenth and seventeenth centuries.  The most important arena (apart from Holland) is England, because it is in England, with its new geographical position as the entrepôt between Europe and America, its achievement of internal economic unity two centuries before France and two and a half centuries before Germany, its constitutional revolution, and its powerful *bourgeoisie* of bankers, shipowners, and merchants, that the transformation of the structure of society is earliest, swiftest, and most complete.  Its essence is the secularization of social and economic philosophy.  The synthesis is resolved into its elements—politics, business, and spiritual exercises; each assumes a separate and independent vitality and obeys the laws of its own being.  The social functions matured within the Church, and long identified with it, are transferred to the State, which in turn is idolized as the dispenser of prosperity and the guardian of civilization.  The theory of a hierarchy of values, embracing all human interests and activities in a system of which the apex is religion, is replaced by the conception of separate and parallel compartments, between which a due balance should be maintained, but which have no vital connection with each other.

The intellectual movement is, of course, very gradual, and is compatible with both throw-backs and precocities which seem to refute its general character.  It is easy to detect premonitions of the coming philosophy in the later Middle Ages, and reversions to an earlier manner at the very end of the seventeenth century. Oresme in the fourteenth century can anticipate the monetary theory associated with the name of Gresham; in the fifteenth century Laurentius de Rudolfis can distinguish between trade bills and finance bills, and St. Antonino describe the significance of capital; while Baxter in 1673 can write a *Christian Directory* in the style of a medieval *Summa,* and Bunyan in 1680 can dissect the economic iniquities of Mr. Badman, who ground the poor with high prices and usury, in the manner of a medieval friar.[7] But the distance traversed in the two centuries between 1500 and 1700 is, nevertheless, immense. At the earlier date, though economic rationalism has proceeded far in Italy, the typical economic systems are those of the Schoolmen; the typical popular teaching is that of the sermon, or of manuals such as *Dives et Pauper;* the typical appeal in difficult cases of conscience is to the Bible, the Fathers, the canon law and its interpreters; the typical controversy is carried on in terms of morality and religion as regularly and inevitably as two centuries later it is conducted in terms of economic expediency.

It is not necessary to point out that the age of Henry VIII and Thomas Cromwell had nothing to learn from the twentieth century as to the niceties of political intrigue or commercial sharp practice. But a cynical unscrupulousness in high places is not incompatible with a general belief in the validity of moral standards which are contradicted by it. No one can read the discussions which took place between 1500 and 1550 on three burning issues—the rise in prices, capital and interest, and the land question in England—without being struck by the constant appeal from the new and clamorous economic interests of the day to the traditional Christian morality, which in social organization, as in the relations of individuals, is still conceived to be the final authority. It is because it is regarded as the final authority that the officers of the Church claim to be heard on questions of social policy, and that, however Catholics, Anglicans, Lutherans, and Calvinists may differ on doctrine or ecclesiastical government, Luther and Calvin, Latimer and Laud, John Knox and the Pilgrim Fathers are agreed that social morality is the province of the Church, and are prepared both to teach it, and to enforce it, when necessary, by suitable discipline.

By the middle of the seventeenth century all that is altered. After the Restoration, we are in a new world of economic, as well as of political, thought. The claim of religion, at best a shadowy claim, to maintain rules of good conscience in economic affairs finally vanished with the destruction of Laud's experiment in a confessional State, and with the failure of the work of the Westminster Assembly. After the Civil War, the attempt to maintain the theory that there was a Christian standard of economic conduct was impossible, not only because of lay opposition, but because the division of the Churches made it evident that no common standard existed which could be enforced by ecclesiastical machinery. The doctrine of the Restoration economists,[8] that, as proved by the experience of Holland, trade and tolerance flourished together, had its practical significance in the fact that neither could prosper without large concessions to individualism.

The ground which is vacated by the Christian moralist is quickly occupied by theorists of another order. The future for the next two hundred years is not with the attempt to reaffirm, with due allowance for altered circumstances, the conception that a moral rule is binding on Christians in their economic transactions, but with the new science of Political Arithmetic, which asserts, at first with hesitation and then with confidence, that no moral rule beyond the letter of the law exists. Influenced in its method by the contemporary progress of mathematics and physics, it handles economic phenomena, not as a casuist, concerned to distinguish right from wrong, but as a scientist, applying a new calculus to impersonal economic forces. Its method, temper, and assumptions are accepted by all educated men, including the clergy, even though its particular conclusions continue for long to be disputed. Its greatest English exponent, before the days of Adam Smith, is the Reverend Dr. Tucker, Dean of Gloucester.

Some of the particular stages in this transition will be discussed later. But that there was a transition, and that the intellectual and moral conversion which it produced was not less momentous than the effect of some more familiar intellectual revolutions, is undeniable. Nor is it to be refuted by insisting that economic motives and economic needs are as old as history, or that the appeal to religion is often a decorous drapery for a triumphant materialism. A medieval cynic, in expounding the canon law as to usury, remarked that "he who takes it goes to hell, and he who does not goes to the workhouse." [9] Mr. Coulton does well to remind us that, even in the Age of Faith, resounding principles were compatible with very sordid practice. In a discussion which has as its subject social thought, not the history of business organization, it is not necessary to elaborate that truism. Only the credulous or the disillusioned will contrast successive periods as light with darkness or darkness with light, or yield to the temper which finds romantic virtues in every age except its own. To appraise the merits of different theories of social organization must be left to those who feel confident that they possess an adequate criterion. All that can be attempted in these pages is to endeavor to understand a few among them.

For, after all, because doctrine and conduct diverge, it does not follow that to examine the former is to hunt abstractions. That men should have thought as they did is sometimes as significant as that they should have acted as they did, and not least significant when thought and practice are at variance.  It may be true that "theory is a criticism of life only in the same sense as a good man is a criticism of a bad one."  But the emphasis of the theorist on certain aspects and values is not arbitrary, but is itself an interpretation, and, if his answers are to be discounted, his questions are none the less evidence as to the assumptions of the period in which they were asked.  It would be paradoxical to dismiss Machiavelli and Locke and Smith and Bentham as irrelevant to the political practice of their age, merely on the ground that mankind has still to wait for the ideal Prince or Whig or Individualist or Utilitarian.  It is not less paradoxical to dismiss those who formulated economic and social theories in the Middle Ages or in the sixteenth century merely because, behind canon law and *summæ* and sermons, behind the good ordinances of borough and gild, behind statutes and proclamations and prerogative courts, there lurked the immutable appetites of the economic man.

There is an evolution of ideas, as well as of organisms, and the quality of civilization depends, as Professor Wallas has so convincingly shown, on the transmission, less of physical qualities, than of a complex structure of habits, knowledge, and beliefs, the destruction of which would be followed within a year by the death of half the human race. Granted that the groundwork of inherited dispositions with which the individual is born has altered little in recorded history, the interests and values which compose his world have undergone a succession of revolutions.  The conventional statement that human nature does not change is plausible only so long as attention is focused on those aspects of it which are least distinctively human.  The wolf is today what he was when he was hunted by Nimrod.  But, while men are born with many of the characteristics of wolves, man is a wolf domesticated, who both transmits the arts by which he has been partially tamed and improves upon them. He steps into a social inheritance, to which each generation adds its own contribution of good and evil, before it bequeaths it to its successors.

There is a moral and religious, as well as a material, environment, which sets its stamp on the individual, even when he is least conscious of it. And the effect of changes in this environment is not less profound. The economic categories of modern society, such as property, freedom of contract and competition, are as much a part of its intellectual furniture as its political conceptions, and, together with religion, have probably been the most potent force in giving it its character. Between the conception of society as a community of unequal classes with varying functions, organized for a common end, and that which regards it as a mechanism adjusting itself through the play of economic motives to the supply of economic needs; between the idea that a man must not take advantage of his neighbor's necessity, and the doctrine that "man's self-love is God's providence"; between the attitude which appeals to a religious standard to repress economic appetites, and that which regards expediency as the final criterion—there is a chasm which no theory of the permanence and ubiquity of economic interests can bridge, and which deserves at least to be explored. To examine how the latter grew out of the former; to trace the change, from a view of economic activity which regarded it as one among other kinds of moral conduct, to the view of it as dependent upon impersonal and almost automatic forces; to observe the struggle of individualism, in the face of restrictions imposed in the name of religion by the Church and of public policy by the State, first denounced, then palliated, then triumphantly justified in the name of economic liberty; to watch how ecclesiastical authority strives to maintain its hold upon the spheres it had claimed and finally abdicates them—to do this is not to indulge a vain curiosity, but to stand at the sources of rivulets which are now a flood.

Has religious opinion in the past regarded questions of social organization and economic conduct as irrelevant to the life of the spirit, or has it endeavored not only to christianize the individual but to make a Christian civilization? Can religion admit the existence of a sharp antithesis between personal morality and the practices which are permissible in business? Does the idea of a Church involve the acceptance of any particular standard of social ethics, and, if so, ought a Church to endeavor to enforce it as among the obligations incumbent on its members? Such are a few of the questions which men are asking today, and on which a more competent examination of history than I can hope to offer might throw at any rate an oblique and wavering light.

### I. THE SOCIAL ORGANISM

We are asking these questions today. Men were asking the same questions, though in different language, throughout the sixteenth century. It is a commonplace that modern economic history begins with a series of revolutionary changes in the direction and organization of commerce, in finance, in prices, and in agriculture. To the new economic situation men brought a body of doctrine, law and tradition, hammered out during the preceding three centuries. Since the new forces were bewildering, and often shocking, to conservative consciences, moralists and religious teachers met them at first by a re-affirmation of the traditional doctrines, by which, it seemed, their excesses might be restrained and their abuses corrected. As the changed environment became, not a novelty, but an established fact, these doctrines had to be modified. As the effects of the Reformation developed, different churches produced characteristic differences of social opinion.

But these were later developments, which only gradually became apparent. The new economic world was not accepted without a struggle. Apart from a few extremists, the first generation of reformers were rarely innovators in matters of social theory, and quoted Fathers and church councils, decretals and canon lawyers, in complete unconsciousness that innovations in doctrine and church government involved any breach with what they had learned to regard as the moral tradition of Christendom. Hence the sixteenth century sees a collision, not only between different schools of religious thought, but between the changed economic environment and the accepted theory of society. To understand it, one must place oneself .at the point from which it started. One must examine, however summarily, the historical background.

That background consisted of the body of social theory, stated and implicit, which was the legacy of the Middle Ages. The formal teaching was derived from the Bible, the works of the Fathers and Schoolmen, the canon law and its commentators, and had been popularized in sermons and religious manuals. The informal assumptions were those implicit in law, custom, and social institutions. Both were complex, and to speak of them as a unity is to sacrifice truth to convenience. It may be that the political historian is justified when he covers with a single phrase the five centuries or more to which tradition has assigned the title of the Middle Ages. For the student of economic conditions that suggestion of homogeneity is the first illusion to be discarded.

The medieval economic world was marked, it is true, by certain common characteristics. They sprang from the fact that on the west it was a closed system, that on the north it had so much elbow-room as was given by the Baltic and the rivers emptying themselves into it, and that on the east, where it was open, the apertures were concentrated along a comparatively short coast-line from Alexandria to the Black Sea, so that they were easily commanded by any naval power dominating the eastern Mediterranean, and easily cut by any military power which could squat across the trade routes before they reached the sea. While, however, these broad facts determined that the two main currents of trade should run from east to west and north to south, and that the most progressive economic life of the age should cluster in the regions from which these currents started and where they met, within this general economic framework there was the greatest variety of condition and development. The contours of economic civilization ran on different lines from those of subsequent centuries, but the contrast between mountain and valley was not less clearly marked. If the sites on which a complex economic structure rose were far removed from those of later generations, it flourished none the less where conditions favored its growth. In spite of the ubiquity of manor and gild, there was as much difference between the life of a center of capitalist industry, like fifteenth-century Flanders, or a center of capitalist finance, like fifteenth-century Florence, and a pastoral society exporting raw materials and a little food, like medieval England, as there is between modern Lancashire or London and modern Denmark. To draw from English conditions a picture of a whole world stagnating in economic squalor, or basking in economic innocence, is as absurd as to reconstruct the economic life of Europe in the twentieth century from a study of the Shetland Islands or the Ukraine. The elements in the social theory of the Middle Ages were equally various, and equally changing. Even if the student confines himself to the body of doctrine which is definitely associated with religion, and takes as typical of it the *Summæ* of the Schoolmen, he finds it in

constant process of development. The economic teaching of
St. Antonino in the fifteenth century, for example, was far
more complex and realistic than that of St. Thomas in the
thirteenth, and down to the very end of the Middle Ages
the best-established and most characteristic parts of the system—for example, the theory of prices and of usury—so
far from being stationary, were steadily modified and elaborated.

There are, perhaps, four main attitudes which religious
opinion may adopt toward the world of social institutions
and economic relations. It may stand on one side in ascetic
aloofness and regard them as in their very nature the sphere
of unrighteousness, from which men *may* escape—from
which, if they consider their souls, they *will* escape—but
which they can conquer only by flight. It may take them
for granted and ignore them, as matters of indifference belonging to a world with which religion has no concern; in
all ages the prudence of looking problems boldly in the face
and passing on has seemed too self-evident to require justification. It may throw itself into an agitation for some particular reform, for the removal of some crying scandal, for
the promotion of some final revolution, which will inaug-
urate the reign of righteousness on earth. It may at once
accept and criticize, tolerate and amend, welcome the gross
world of human appetites, as the squalid scaffolding from
amid which the life of the spirit must rise, and insist that
this also is the material of the Kingdom of God. To such
a temper, all activities divorced from religion are brutal or
dead, but none are too mean to be beneath or too great to
be above it, since all, in their different degrees, are touched
with the spirit which permeates the whole. It finds its most
sublime expression in the words of Piccarda: "Paradise is
everywhere, though the grace of the highest good is not
shed everywhere in the same degree."

Each of these attitudes meets us today. Each meets us

in the thought of the Middle Ages, as differences of period and place and economic environment and personal temperament evoke it. In the early Middle Ages the ascetic temper predominates. Lanfranc, for example, who sees nothing in economic life but the struggle of wolves over carrion, thinks that men of business can hardly be saved, for they live by cheating and profiteering.[10] It is monasticism, with its repudiation of the prizes and temptations of the secular world, which is *par excellence* the life of religion. As one phase of it succumbed to ease and affluence, another rose to restore the primitive austerity, and the return to evangelical poverty, preached by St. Francis but abandoned by many of his followers, was the note of the majority of movements for reform. As for indifferentism—what else, for all its communistic phrases, is Wyclif's teaching, that the "just man is already lord of all" and that "in this world God must serve the devil," but an anticipation of the doctrine of celestial happiness as the compensation for earthly misery, to which Hobbes gave a cynical immortality when he wrote that the persecuted, instead of rebelling, "must expect their reward in Heaven," and which Mr. and Mrs. Hammond have revealed as an opiate dulling both the pain and the agitation of the Industrial Revolution? If obscure sects like the Poor Men of Lyons are too unorthodox to be cited, the Friars are not, and it was not only Langland and that gentlemanly journalist, Froissart, who accused them— the phrase has a long history—of stirring up class hatred.

To select from so immense a sea of ideas about society and religion only the specimens that fit the meshes of one's own small net, and to label them "medieval thought," is to beg all questions. Ideas have a pedigree which, if realized, would often embarrass their exponents. The day has long since passed when it could be suggested that only one-half of modern Christianity has its root in medieval religion. There is a medieval Puritanism and rationalism as well as a medieval Catholicism. In the field of ecclesiastical theory, as Mr. Manning has pointed out in his excellent book,[11] Gregory VII and Boniface VIII have their true successors in Calvin and Knox. What is true of religion and political thought is equally true of economic and social doctrines. The social theories of Luther and Latimer, of Bucer and Bullinger, of sixteenth-century Anabaptists and seventeenth-century Levellers, of Puritans like Baxter, Anglicans like Laud, Baptists like Bunyan, Quakers like Bellers, are all the children of medieval parents. Like the Church today in regions which have not yet emerged from savagery, the Church of the earlier Middle Ages had been engaged in an immense missionary effort, in which, as it struggled with the surrounding barbarism, the work of conversion and of social construction had been almost indistinguishable. By the very nature of its task, as much as by the intention of its rulers, it had become the greatest of political institutions. For good or evil it aspired to be, not a sect, but a civilization, and, when its unity was shattered at the Reformation, the different Churches which emerged from it endeavored, according to their different opportunities, to perpetuate the same tradition. Asceticism or renunciation, quietism or indifferentism, the zeal which does well to be angry, the temper which seeks a synthesis of the external order and the religion of the spirit—all alike, in one form or another, are represented in the religious thought and practice of the Middle Ages.

All are represented in it, but not all are equally representative of it. Of the four attitudes suggested above, it is the last which is most characteristic. The first fundamental assumption which is taken over by the sixteenth century is that the ultimate standard of human institutions and activities is religion. The architectonics of the system had been worked out in the *Summæ* of the Schoolmen. In sharp contrast to the modern temper, which takes the destination for granted, and is thrilled by the hum of the engine, medieval religious thought strains every interest and activity, by however arbitrary a compression, into the service of a single idea. The lines of its scheme run up and down, and, since purpose is universal and all-embracing, there is, at least in theory, no room for eccentric bodies which move in their own private orbit. That purpose is set by the divine plan of the universe. "The perfect happiness of man cannot be other than the vision of the divine essence." [12]

Hence all activities fall within a single system, because all, though with different degrees of immediateness, are related to a single end, and derive their significance from it. The Church in its wider sense is the Christian Commonwealth, within which that end is to be realized; in its narrower sense it is the hierarchy divinely commissioned for its interpretation; in both it embraces the whole of life, and its authority is final. Though practice is perpetually at variance with theory, there is no absolute division between the inner and personal life, which is "the sphere of religion," and the practical interests, the external order, the impersonal mechanism, to which, if some modern teachers may be trusted, religion is irrelevant.

There is no absolute division, but there is a division of quality. There are—to use a modern phrase—degrees of reality. The distinctive feature of medieval thought is that contrasts which later were to be presented as irreconcilable antitheses appear in it as differences within a larger unity, and that the world of social organization, originating in physical necessities, passes by insensible gradations into that of the spirit. Man shares with other animals the necessity of maintaining and perpetuating his species; in addition, as a natural creature, he has what is peculiar to himself, an inclination to the life of the intellect and of society— "to know the truth about God and to live in communities." [13] These activities, which form his life according to the law of nature, may be regarded, and sometimes are regarded, as indifferent or hostile to the life of the spirit. But the characteristic thought is different. It is that of a synthesis.

The contrast between nature and grace, between human appetites and interests and religion, is not absolute, but relative. It is a contrast of matter and the spirit informing it, of stages in a process, of preparation and fruition. Grace works on the unregenerate nature of man, not to destroy it, but to transform it. And what is true of the individual is true of society. An attempt is made to give it a new significance by relating it to the purpose of human life as known by revelation. In the words of a famous (or notorious) Bull: "The way of religion is to lead the things which are lower to the things which are higher through the things which are intermediate. According to the law of the universe all things are not reduced to order equally and immediately; but ·the lowest through the intermediate, the intermediate through the higher." [14] Thus social institutions assume a character which may almost be called sacramental, for they are the outward and imperfect expression of a supreme spiritual reality. Ideally conceived, society is an organism of different grades, and human activities form a hierarchy of functions, which differ in kind and in significance, but each of which is of value on its own plane, provided that it is governed, however remotely, by the end which is common to all. Like the celestial order, of which it is the dim reflection, society is stable, because it is straining upwards:

> Anzi è formale ad esto beato esse
> Tenersi dentro alla divina voglia,
> Per ch' una fansi nostre voglie stesse.

Needless to say, metaphysics, however sublime, were not the daily food of the Middle Ages, any more than of today. The fifteenth century saw an outburst of commercial activity and of economic speculation, and by the middle of it all this teaching was becoming antiquated. Needless to say, also, general ideas cannot be kept in compartments, and the teleology of medieval speculation colored the interpretation of common affairs, as it was colored by physics in the eighteenth century and by the idea of evolution in the nineteenth. If the first legacy of the Middle Ages to the sixteenth century was the idea of religion as embracing all aspects of human life, the second and third flowed naturally from the working of that idea in the economic environment of the time. They may be called, respectively, the functional view of class organization, and the doctrine of economic ethics.

From the twelfth century to the sixteenth, from the work of Beckett's secretary in 1159 to the work of Henry VIII's chaplain in 1537, the analogy by which society is described —an analogy at once fundamental and commonplace—is the same.[15] Invoked in every economic crisis to rebuke extortion and dissension with a high doctrine of social solidarity, it was not finally discarded till the rise of a theoretical individualism in England in the seventeenth century. It is that of the human body. The gross facts of the social order are accepted in all their harshness and brutality. They are accepted with astonishing docility, and, except on rare occasions, there is no question of reconstruction. What they include is no trifle. It is nothing less than the whole edifice of feudal society—class privilege, class oppression, exploitation, serfdom. But these things cannot, it is thought, be treated as simply alien to religion, for religion is all-comprehensive. They must be given some ethical meaning, must be shown to be the expression of some larger plan. The meaning given them is simple. The facts of class status and inequality were rationalized in the Middle Ages by a functional theory of society, as the facts of competition were rationalized in the eighteenth by the theory of economic harmonies;

and the former took the same delight in contemplating the moral purpose revealed in social organization as the latter in proving that to the curious mechanism of human society a moral purpose was superfluous or disturbing. Society, like the human body, is an organism composed of different members. Each member has its own function, prayer, or defense, or merchandise, or tilling the soil. Each must receive the means suited to its station, and must claim no more. Within classes there must be equality; if one takes into his hand the living of two, his neighbor will go short. Between classes there must be inequality; for otherwise a class cannot perform its function, or—a strange thought to us—enjoy its rights. Peasants must not encroach on those above them. Lords must not despoil peasants. Craftsmen and merchants must receive what will maintain them in their calling, and no more.

As a rule of social policy, the doctrine was at once repressive and protective. "There is degree above degree, as reason is, and skill it is that men do their devoir thereas it is due. But certes, extortions and despite of your underlings is damnable." [16] As a philosophy of society, it attempted to spiritualize the material by incorporating it in a divine universe, which should absorb and transform it. To that process of transmutation the life of mere money-making was recalcitrant, and hence, indeed, the stigma attached to it. For, in spite of the ingenuity of theorists, finance and trade, the essense of which seemed to be, not service, but a mere *appetitus divitiarum infinitus,* were not easily interpreted in terms of social function. Comparatively late intruders in a world dominated by conceptions hammered out in a pre-commercial age, they were never fitted harmoniously into the medieval synthesis, and ultimately, when they grew to their full stature, were to contribute to its overthrow. But the property of the feudal lord, the labor of the peasant or the craftsman, even the ferocity of the warrior, were not dismissed as hostile or indifferent to the life of the spirit. Touched by the spear of Ithuriel, they were to be sublimated into service, vocation and chivalry, and the ritual which surrounded them was designed to emphasize that they had undergone a re-dedication at the hands of religion. Baptized by the Church, privilege and power became office and duty.

That the reconciliation was superficial, and that in attempting it the Church often degraded itself without raising the world, is as indisputable as that its tendency was to dignify material interests, by stamping them with the impress of a universal design. Gentlemen took hard tallages and oppressed the poor; but it was something that they should be told that their true function was "to defend God's law by power of the world." [17] Craftsmen—the burden of endless sermons—worked deceitfully; but it was perhaps not wholly without value that they should pay even lip-service to the ideal of so conducting their trade, that the common people should not be defrauded by the evil ingenuity of those exercising the craft. If lord and peasant, merchant and artisan, burgess and villager, pressed each other hard, was it meaningless to meet their struggles with an assertion of universal solidarity, to which economic convenience and economic power must alike give way? "The health of the whole commonwealth will be assured and vigorous, if the higher members consider the lower and the lower answer in like manner the higher, so that each is in its turn a member of every other." [18]

If the medieval moralist was often too naïve in expecting sound practice as the result of lofty principles alone, he was at least free from that not unfashionable form of credulity which expects it from their absence or from their opposite. To say that the men to whom such teaching was addressed went out to rob and cheat is to say no more than that they were men. Nor is it self-evident that they would have been more likely to be honest, if they had been informed, like some of their descendants, that competition was designed by Providence to provide an automatic substitute for honesty. Society was interpreted, in short, not as the expression of economic self-interest, but as held together by a system of mutual, though varying, obligations. Social well-being exists, it was thought, in so far as each class performs its functions and enjoys the rights proportioned thereto. "The Church is divided in these three parts, preachers, and defenders, and . . . laborers. . . . As she is our mother, so she is a body, and health of this body stands in this, that one part of her answer to another, after the same measure that Jesus Christ has ordained it. . . . Kindly man's hand helps his head, and his eye helps his foot, and his foot his body . . . and thus should it be in parts of the Church. . . . As divers parts of man served unkindly to man if one took the service of another and left his own proper work, so divers parts of the Church have proper works to serve God; and if one part leave his work that God has limited him and take work of another part, sinful wonder is in the Church. . . . Surely the Church shall never be whole before proportions of her parts be brought again by this heavenly leech and [by] medicine of men." [19]

Speculation does not develop *in vacuo*. It echoes, however radical it is, the established order. Clearly this patriarchal doctrine is a softened reflection of the feudal land system. Not less clearly the Church's doctrine of economic ethics is the expression of the conditions of medieval industry. A religious philosophy, unless it is frankly to abandon nine-tenths of conduct to the powers of darkness, cannot admit the doctrine of a world of business and economic relations self-sufficient and divorced from ethics and religion. But the facts may be difficult to moralize, or they may be relatively easy. Over a great part of Europe in the later Middle Ages, the economic environment was less intractable than it had been in the days of the Empire or than it is today. In the great commercial centers there was sometimes, it is true, a capitalism as inhuman as any which the world has seen, and from time to time ferocious class wars between artisans and merchants.[20] But outside them trade, industry, the money market, all that we call the economic system, was not a system, but a mass of individual trades and individual dealings. Pecuniary transactions were a fringe on a world of natural economy. There was little mobility or competition. There was very little large-scale organization. With some important exceptions, such as the textile workers of Flanders and Italy, who, in the fourteenth century, again and again rose in revolt, the medieval artisan, especially in backward countries like England, was a small master. The formation of temporary organizations, or "parliaments," of wage-earners, which goes on in London even before the end of the thirteenth century,[21] and the growth of journeymen's associations in the later Middle Ages, are a proof that the conditions which produced modern trade unionism were not unknown. But even in a great city like Paris the 128 gilds which existed at the end of the thirteenth century appear to have included 5,000 masters, who employed not more than 6,000 to 7,000 journeymen. At Frankfurt-am-Main in 1387 actually not more than 750 to 800 journeymen are estimated to have been in the service of 1,554 masters.[22]

In cities of this kind, with their freedom, their comparative peace, and their strong corporate feeling, large enough to be prolific of associations and small enough for each man to know his neighbor, an ethic of mutual aid was not wholly impossible, and it is in the light of such conditions that the most characteristic of medieval industrial institutions is to be interpreted. To suggest that anything like a majority of medieval workers were ever members of a craft gild is extravagant. In England, at any rate, more than nine-tenths were peasants, among whom, though friendly societies called gilds were common, there was naturally no question of craft organization. Even in the towns it is a question whether there was not a considerable population of casual workers—consider only the number of unskilled workers that must have been required as laborers by the craftsmen building a cathedral in the days before mechanical cranes—who were rarely organized in permanent societies. To invest the craft gilds with a halo of economic chivalry is not less inappropriate. They were, first and foremost, monopolists, and the cases in which their vested interests came into collision with the consumer were not a few. Wyclif, with his almost modern devotion to the conception of a unitary society over-riding particular interests for the common good, was naturally prejudiced against corporations, on the ground that they distracted social unity by the intrusion of sectarian cupidities and sinister ambitions; but there was probably from time to time more than a little justification for his complaint that "all new fraternities or gilds made of men seem openly to run in this curse [against false conspirators]," because "they conspire to bear up each other, yea, in wrong, and oppress other men in their right by their wit and power." [23]  It is significant that the most striking of the projects of political and social reconstruction produced in Germany in the century before the Reformation proposed the complete abolition of gilds, as intolerably corrupt and tyrannical.[24]

There are, however, monopolists and monopolists. An age in which combinations are not tempted to pay lip-service to religion may do well to remember that the characteristic, after all, of the medieval gild was that, if it sprang from economic needs, it claimed, at least, to subordinate them to social interests, as conceived by men for whom the social and the spiritual were inextricably intertwined. "Tout ce petit monde antique," writes the historian of French gilds, "était fortement imbu des idées chrétiennes sur le juste salaire et le juste prix; sans doute il y avait alors, comme aujourd'hui, des cupidités et des convoitises; mais une règle puissante s'imposait à tous et d'une manière générale exigeait pour chacun le pain quotidien promis par l'Evangile." [25] The attempt to preserve a rough equality among "the good men of the mistery," to check economic egotism by insisting that every brother shall share his good fortune with another and stand by his neighbor in need, to resist the encroachments of a conscienceless money-power, to preserve professional standards of training and craftsmanship, and to repress by a strict corporate discipline the natural appetite of each to snatch special advantages for himself to the injury of all—whether these things outweigh the evils of conservative methods and corporate exclusiveness is a question which each student will answer in accordance with his own predilections. What is clear, at least, is that both the rules of fraternities and the economic teaching of the Church were prompted by the problems of a common environment. Much that is now mechanical was then personal, intimate and direct, and there was little room for organization on a scale too vast for the standards that are applied to individuals, or for the doctrine which silences scruples and closes all accounts with the final plea of economic expediency.

Such an environment, with its personal economic relations, was a not unfavorable field for a system of social ethics. And the Church, which brought to its task the tremendous claim to mediate between even the humblest activity and the divine purpose, sought to supply it. True, its teaching was violated in practice, and violated grossly, in the very citadel of Christendom which promulgated it. Contemporaries were under no illusion as to the reality of economic motives in the Age of Faith. They had only to look at Rome. From the middle of the thirteenth century a

continuous wail arises against the iniquity of the Church,
and its burden may be summed up in one word, "avarice."
At Rome, everything is for sale. What is reverenced is the
gospel, not according to St. Mark, but according to the
marks of silver.[26]

> Cum ad papam veneris, habe pro constanti,
> Non est locus pauperi, soli favet danti.
>
> .    .    .    .    .
>
> Papa, si rem tangimus, nomen habet a re,
> Quicquid habent alii, solus vult papare;
> Vel, si verbum gallicum vis apocopare,
> *'Payez, payez,' dit le mot,* si vis impetrare.[27]

The Papacy might denounce usurers, but, as the center
of the most highly organized administrative system of the
age, receiving remittances from all over Europe, and receiving them in money at a time when the revenue of other
Governments still included personal services and payments
in kind, it could not dispense with them. Dante put the
Cahorsine money-lenders in hell, but a Pope gave them the
title of "peculiar sons of the Roman Church." [28] Grosstête
rebuked the Lombard bankers, and a bishop of London expelled them, but papal protection brought them back.[29]
Archbishop Peckham, a few years later, had to implore
Pope Nicholas III to withdraw a threat of excommunication, intended to compel him to pay the usurious interest
demanded by Italian money-lenders, though, as the archbishop justly observed, "by your Holiness's special man-
date, it would be my duty to take strong measures against
such lenders." [30] The Papacy was, in a sense, the greatest
financial institution of the Middle Ages, and, as its fiscal
system was elaborated, things became, not better, but worse.
The abuses which were a trickle in the thirteenth century
were a torrent in the fifteenth. And the frailties of Rome,
if exceptional in their notoriety, can hardly be regarded as

unique. Priests, it is from time to time complained, engage in trade and take usury.[31] Cathedral chapters lend money at high rates of interest. The profits of usury, like those of simony, should have been refused by churchmen, as hateful to God; but a bishop of Paris, when consulted by a usurer as to the salvation of his soul, instead of urging restitution, recommended him to dedicate his ill-gotten wealth to the building of Notre-Dame.[32] "Thus," exclaimed St. Bernard, as he gazed at the glories of Gothic architecture, "wealth is drawn up by ropes of wealth, thus money bringeth money. . . . O vanity of vanities, yet no more vain than insane! The Church is resplendent in her walls, beggarly in her poor. She clothes her stones in gold, and leaves her sons naked." [33]

The picture is horrifying, and one must be grateful to those, like M. Luchaire and Mr. Coulton, who demolish romance. But the denunciation of vices implies that they are recognized as vicious; to ignore their condemnation is not less one-sided than to conceal their existence; and, when the halo has vanished from practice, it remains to ask what principles men valued, and what standards they erected. The economic doctrines elaborated in the *Summæ* of the Schoolmen, in which that question receives its most systematic answer, have not infrequently been dismissed as the fanciful extravagances of writers disqualified from throwing light on the affairs of this world by their morbid preoccupation with those of the next. In reality, whatever may be thought of their conclusions, both the occasion and the purpose of scholastic speculations upon economic questions were eminently practical. The movement which prompted them was the growth of trade, of town life, and of a commercial economy, in a world whose social categories were still those of the self-sufficing village and the feudal hierarchy. The object of their authors was to solve the problems to which such developments gave rise. It was to reconcile the new contractual relations, which sprang from economic expansion, with the traditional morality expounded by the Church. Viewed by posterity as reactionaries, who damned the currents of economic enterprise with an irrelevant appeal to Scripture and to the Fathers, in their own age they were the pioneers of a liberal intellectual movement. By lifting the weight of antiquated formulæ they cleared a space within the stiff framework of religious authority for new and mobile economic interests, and thus supplied an intellectual justification for developments which earlier generations would have condemned.

The mercantilist thought of later centuries owed a considerable debt to scholastic discussions of money, prices, and interest. But the specific contributions of medieval writers to the technique of economic theory were less significant than their premises. Their fundamental assumptions, both of which were to leave a deep imprint on the social thought of the sixteenth and seventeenth centuries, were two: that economic interests are subordinate to the real business of life, which is salvation, and that economic conduct is one aspect of personal conduct, upon which, as on other parts of it, the rules of morality are binding. Material riches are necessary; they have a secondary importance, since without them men cannot support themselves and help one another; the wise ruler, as St. Thomas said,[34] will consider in founding his State the natural resources of the country. But economic motives are suspect. Because they are powerful appetites, men fear them, but they are not mean enough to applaud them. Like other strong passions, what they need, it is thought, is not a clear field, but repression. There is no place in medieval theory for economic activity which is not related to a moral end, and to found a science of society upon the assumption that the appetite for economic gain is a constant and measurable force, to be accepted, like other natural forces, as an inevitable and self-evident *datum* would have appeared to the medieval thinker as hardly less irrational or less immoral than to make the premise of social philosophy the unrestrained operation of such necessary human attributes as pugnacity or the sexual instinct. The outer is ordained for the sake of the inner; economic goods are instrumental—*sicut quædam adminicula, quibus adjuvamur ad tendendum in beatitudinem.* "It is lawful to desire temporal blessings, not putting them in the first place, as though setting up our rest in them, but regarding them as aids to blessedness, inasmuch as they support our corporal life and serve as instruments for acts of virtue." [35] Riches, as St. Antonino says, exist for man, not man for riches.

At every turn, therefore, there are limits, restrictions, warnings against allowing economic interests to interfere with serious affairs. It is right for a man to seek such wealth as is necessary for a livelihood in his station. To seek more is not enterprise, but avarice, and avarice is a deadly sin. Trade is legitimate; the different resources of different countries show that it was intended by Providence. But it is a dangerous business. A man must be sure that he carries it on for the public benefit, and that the profits which he takes are no more than the wages of his labor. Private property is a necessary institution, at least in a fallen world; men work more and dispute less when goods are private than when they are common. But it is to be tolerated as a concession to human frailty, not applauded as desirable in itself; the ideal—if only man's nature could rise to it—is communism. "Communis enim," wrote Gratian in his *decretum,* "usus omnium, quae sunt in hoc mundo, omnibus hominibus esse debuit." [36] At best, indeed, the estate is somewhat encumbered. It must be legitimately acquired. It must be in the largest possible number of hands. It must provide for the support of the poor. Its use must as far as practicable be common. Its owners must be ready to share it with those who need, even if they are not in actual destitution.    Such were the conditions which commended themselves to an archbishop of the business capital of fifteenth-century Europe.[37]    There have been ages in which they would have been described, not as a justification of property, but as a revolutionary assault on it.    For to defend the property of the peasant and small master is necessarily to attack that of the monopolist and usurer, which grows by devouring it.

The assumption on which all this body of doctrine rested was simple.    It was that the danger of economic interests increased in direct proportion to the prominence of the pecuniary motives associated with them.    Labor—the common lot of mankind—is necessary and honorable; trade is necessary, but perilous to the soul; finance, if not immoral, is at best sordid and at worst disreputable.    This curious inversion of the social values of more enlightened ages is best revealed in medieval discussions of the ethics of commerce. The severely qualified tolerance extended to the trader was partly, no doubt, a literary convention derived from classical models; it was natural that Aquinas should laud the State which had small need of merchants because it could meet its needs from the produce of its own soil; had not the Philosopher himself praised $αὐταρχεία$ ?    But it was a convention which coincided with a vital element in medieval social theory, and struck a responsive note in wide sections of medieval society.    It is not disputed, of course, that trade is indispensable; the merchant supplements the deficiencies of one country with the abundance of another.    If there were no private traders, argued Duns Scotus, whose indulgence was less carefully guarded, the governor would have to engage them.    Their profits, therefore, are legitimate, and they may include, not only the livelihood appropriate to the trader's status, but payment for labor, skill, and risk.[38]

The defence, if adequate, was somewhat embarrassing.

For why should a defence be required? The insistence that trade is not positively sinful conveys a hint that the practices of traders may be, at least, of dubious propriety. And so, in the eyes of most medieval thinkers, they are. *Summe periculosa est venditionis et emptionis negotiatio.*[39] The explanation of that attitude lay partly in the facts of contemporary economic organization. The economy of the medieval borough—consider only its treatment of food supplies and prices—was one in which consumption held somewhat the same primacy in the public mind, as the undisputed arbiter of economic effort, as the nineteenth century attached to profits. The merchant pure and simple, though convenient to the Crown, for whom he collected taxes and provided loans, and to great establishments such as monasteries, whose wool he bought in bulk, enjoyed the double unpopularity of an alien and a parasite. The best practical commentary on the tepid indulgence extended by theorists to the trader is the network of restrictions with which medieval policy surrounded his activities, the recurrent storms of public indignation against him, and the ruthlessness with which boroughs suppressed the middleman who intervened between consumer and producer.

Apart, however, from the color which it took from its environment, medieval social theory had reasons of its own for holding that business, as distinct from labor, required some special justification. The suspicion of economic motives had been one of the earliest elements in the social teaching of the Church, and was to survive till Calvinism endowed the life of economic enterprise with a new sanctification. In medieval philosophy the ascetic tradition, which condemned all commerce as the sphere of iniquity, was softened by a recognition of practical necessities, but it was not obliterated; and, if reluctant to condemn, it was insistent to warn. For it was of the essence of trade to drag into a position of solitary prominence the acquisitive appetites; and towards those appetites, which to most modern thinkers have seemed the one sure social dynamic, the attitude of the medieval theorist was that of one who holds a wolf by the ears.   The craftsman labors for his living; he seeks what is sufficient to support him, and no more.   The merchant aims, not merely at livelihood, but at profit.   The traditional distinction was expressed in the words of Gratian: "Whosoever buys a thing, not that he may sell it whole and unchanged, but that it may be a material for fashioning something, he is no merchant.   But the man who buys it in order that he may gain by selling it again unchanged and as he bought it, that man is of the buyers and sellers who are cast forth from God's temple." [40]   By very definition a man who "buys in order that he may sell dearer," the trader is moved by an inhuman concentration on his own pecuniary interest, unsoftened by any tincture of public spirit or private charity.   He turns what should be a means into an end, and his occupation, therefore, "is justly condemned, since, regarded in itself, it serves the lust of gain." [41]

The dilemma presented by a form of enterprise at once perilous to the soul and essential to society was revealed in the solution most commonly propounded for it.   It was to treat profits as a particular case of wages, with the qualification that gains in excess of a reasonable remuneration for the merchant's labor were, though not illegal, reprehensible as *turpe lucrum*.   The condition of the trader's exoneration is that "he seeks gain, not as an end, but as the wages of his labor." [42]   Theoretically convenient, the doctrine was difficult of application, for evidently it implied the acceptance of what the sedate irony of Adam Smith was later to describe as "an affectation not very common among merchants."   But the motives which prompted it were characteristic.   The medieval theorist condemned as a sin precisely that effort to achieve a continuous and unlimited increase in material wealth which modern societies applaud as a quality, and the vices for which he reserved his most merciless denunciations were the more refined and subtle of the economic virtues. "He who has enough to satisfy his wants," wrote a Schoolman of the fourteenth century, "and nevertheless ceaselessly labors to acquire riches, either in order to obtain a higher social position, or that subsequently he may have enough to live without labor, or that his sons may become men of wealth and importance—all such are incited by a damnable avarice, sensuality, or pride." [43]    Two and a half centuries later, in the midst of a revolution in the economic and spiritual environment, Luther, in even more unmeasured language, was to say the same.[44]    The essence of the argument was that payment may properly be demanded by the craftsmen who make the goods, or by the merchants who transport them, for both labor in their vocation and serve the common need.    The unpardonable sin is that of the speculator or the middleman, who snatches private gain by the exploitation of public necessities.    The true descendant of the doctrines of Aquinas is the labor theory of value.    The last of the Schoolmen was Karl Marx.

## II.    THE SIN OF AVARICE

If such ideas were to be more than generalities, they required to be translated into terms of the particular transactions by which trade is conducted and property acquired. Their practical expression was the body of economic casuistry, in which the best-known elements are the teaching with regard to the just price and the prohibition of usury. These doctrines sprang as much from the popular consciousness of the plain facts of the economic situation as from the theorists who expounded them.    The innumerable fables of the usurer who was prematurely carried to hell, or whose money turned to withered leaves in his strong box, or who (as the scrupulous recorder remarks), "about the year 1240," on entering a church to be married, was crushed by a stone figure falling from the porch, which proved by the grace of God to be a carving of another usurer and his money-bags being carried off by the devil, are more illuminating than the refinements of lawyers.[45]

On these matters, as the practice of borough and manor, as well as of national governments, shows, the Church was preaching to the converted, and to dismiss its teaching on economic ethics as the pious rhetoric of professional moralists is to ignore the fact that precisely similar ideas were accepted in circles which could not be suspected of any unnatural squeamishness as to the arts by which men grow rich. The best commentary on ecclesiastical doctrines as to usury and prices is the secular legislation on similar subjects, for, down at least to the middle of the sixteenth century, their leading ideas were reflected in it. Plain men might curse the chicanery of ecclesiastical lawyers, and gilds and boroughs might forbid their members to plead before ecclesiastical courts; but the rules which they themselves made for the conduct of business had more than a flavor of the canon law. Florence was the financial capital of medieval Europe; but even at Florence the secular authorities fined bankers right and left for usury in the middle of the fourteenth century, and, fifty years later, first prohibited credit transactions altogether, and then imported Jews to conduct a business forbidden to Christians.[46] Cologne was one of the greatest of commercial entrepôts; but, when its successful business man came to make his will, he remembered that trade was perilous to the soul and avarice a deadly sin, and offered what atonement he could by directing his sons to make restitution and to follow some less dangerous occupation than that of the merchant.[47] The burgesses of Coventry fought the Prior over a question of common rights for the best part of a century; but the Court Leet of that thriving business city put usury on a par with adultery and fornication, and decreed that no usurer could become mayor, councillor, or master of the gild.[48]   It was not that laymen were unnaturally righteous; it was not that the Church was all-powerful, though its teaching wound into men's minds through a hundred channels, and survived as a sentiment long after it was repudiated as a command.   It was that the facts of the economic situation imposed themselves irresistibly on both.   In reality, there was no sharp collision between the doctrine of the Church and the public policy of the world of business—its individual practice was, of course, another matter—because both were formed by the same environment, and accepted the same broad assumptions as to social expediency.

The economic background of it all was very simple.   The medieval consumer—we can sympathize with him today more easily than in 1914—is like a traveller condemned to spend his life at a station hotel.   He occupies a tied house and is at the mercy of the local baker and brewer.   Monopoly is inevitable.   Indeed, a great part of medieval industry is a system of organized monopolies, endowed with a public status, which must be watched with jealous eyes to see that they do not abuse their powers.   It is a society of small masters and peasant farmers.   Wages are not a burning question, for, except in the great industrial centers of Italy and Flanders, the permanent wage-earning class is small.   Usury is, as it is today in similar circumstances. For loans are made largely for consumption, not for production.   The farmer whose harvest fails or whose beasts die, or the artisan who loses money, must have credit, seed-corn, cattle, raw materials, and his distress is the money-lender's opportunity.   Naturally, there is a passionate popular sentiment against the engrosser who holds a town to ransom, the monopolist who brings the livings of many into the hands of one, the money-lender who takes advantage of his neighbor's necessities to get a lien on their land and foreclose. "The usurer would not loan to men these goods, but if he hoped winning, that he loves more than charity. Many other sins be more than this usury, but for this men curse and hate it more than other sin." [49]

No one who examines the cases actually heard by the courts in the later Middle Ages will think that resentment surprising, for they throw a lurid light on the possibilities of commercial immorality.[50] Among the peasants and small masters who composed the mass of the population in medieval England, borrowing and lending were common, and it was with reference to their petty transactions, not to the world of high finance, that the traditional attitude towards the money-lender had been crystallized. It was natural that "Juetta [who] is a usuress and sells at a dearer rate for accommodation," and John the Chaplain, *qui est usurarius maximus*,[51] should be regarded as figures at once too scandalous to be tolerated by their neighbors and too convenient to be altogether suppressed. The Church accepts this popular sentiment, gives it a religious significance, and crystallizes it in a system, in which economic morality is preached from the pulpit, emphasized in the confessional, and enforced, in the last resource, through the courts.

The philosophical basis of it is the conception of natural law. "Every law framed by man bears the character of a law exactly to that extent to which it is derived from the law of nature. But if on any point it is in conflict with the law of nature, it at once ceases to be a law; it is a mere perversion of law." [52] The plausible doctrine of compensations, of the long run, of the self-correcting mechanism, has not yet been invented. The idea of a law of nature—of natural justice which ought to find expression in positive law, but which is not exhausted in it—supplies an ideal standard by which the equity of particular relations can be measured. The most fundamental difference between medieval and modern economic thought consists, indeed, in the fact that, whereas the latter normally refers to economic expediency, however it may be interpreted, for the justification of any particular action, policy, or system of organization, the former starts from the position that there is a moral authority to which considerations of economic expediency must be subordinated.  The practical application of this conception is the attempt to try every transaction by a rule of right, which is largely, though not wholly, independent of the fortuitous combinations of economic circumstances.  No man must ask more than the price fixed, either by public authorities, or, failing that, by common estimation.  True, prices even so will vary with scarcity; for, with all their rigor, theologians are not so impracticable as to rule out the effect of changing supplies.  But they will not vary with individual necessity or individual opportunity.  The bugbear is the man who uses, or even creates, a temporary shortage, the man who makes money out of the turn of the market, the man who, as Wyclif says, *must* be wicked, or he could not have been poor yesterday and rich today.[53]

The formal theory of the just price went, it is true, through a considerable development.  The dominant conception of Aquinas—that prices, though they will vary with the varying conditions of different markets, should correspond with the labor and costs of the producer, as the proper basis of the *communis estimatio,* conformity with which was the safeguard against extortion—was qualified by subsequent writers.  Several Schoolmen of the fourteenth century emphasized the subjective element in the common estimation, insisted that the essence of value was utility, and drew the conclusion that a fair price was most likely to be reached under freedom of contract, since the mere fact that a bargain had been struck showed that both parties were satisfied.[54]  In the fifteenth century St. Antonino, who wrote with a highly developed commercial civilization beneath his eyes, endeavored to effect a synthesis, in which the principle of the traditional doctrine should be observed, while the necessary play should be left to economic motives. After a subtle analysis of the conditions affecting value, he concluded that the fairness of a price could at best be a matter only of "probability and conjecture," since it would vary with places, periods and persons. His practical contribution was to introduce a new elasticity into the whole conception by distinguishing three grades of prices—a *gradus pius, discretus,* and *rigidus*. A seller who exceeded the price fixed by more than 50 per cent. was bound, he argued, to make restitution, and even a smaller departure from it, if deliberate, required atonement in the shape of alms. But accidental lapses were venial, and there was a debatable ground within which prices might move without involving sin.[55]

This conclusion, with its recognition of the impersonal forces of the market, was the natural outcome of the intense economic activity of the later Middle Ages, and evidently contained the seeds of an intellectual revolution. The fact that it should have begun to be expounded as early as the middle of the fourteenth century is a reminder that the economic thought of Schoolmen contained elements much more various and much more modern than is sometimes suggested. But the characteristic doctrine was different. It was that which insisted on the just price as the safeguard against extortion. "To leave the prices of goods at the discretion of the sellers is to give rein to the cupidity which goads almost all of them to seek excessive gain." Prices must be such, and no more than such, as will enable each man to "have the necessaries of life suitable for his station." The most desirable course is that they should be fixed by public officials, after making an enquiry into the supplies available and framing an estimate of the requirements of different classes. Failing that, the individual must fix prices for himself, guided by a consideration of "what he must charge in order to maintain his position, and nourish himself suitably in it, and by a reasonable estimate of his expenditure and labor." [56]    If the latter recommendation was a counsel of perfection, the former was almost a platitude.   It was no more than an energetic mayor would carry out before breakfast.

No man, again, may charge money for a loan.   He may, of course, take the profits of partnership, provided that he takes the partner's risks.   He may buy a rent-charge; for the fruits of the earth are produced by nature, not wrung from man.   He may demand compensation—*interesse*—if he is not repaid the principal at the time stipulated.   He may ask payment corresponding to any loss he incurs or gain he foregoes.   He may purchase an annuity, for the payment is contingent and speculative, not certain.   It is no usury when John Deveneys, who has borrowed £19 16s., binds himself to pay a penalty of £40 in the event of failure to restore the principal, for this is compensation for damages incurred; or when Geoffrey de Eston grants William de Burwode three marks of silver in return for an annual rent of six shillings, for this is the purchase of a rent-charge, not a loan; or when James le Reve of London advances £100 to Robert de Bree of Dublin, merchant, with which to trade for two years in Ireland, for this is a partnership; or when the priory of Worcester sells annuities for a capital sum paid down.[57]   What remained to the end unlawful was that which appears in modern economic text-books as "pure interest"—interest as a fixed payment stipulated in advance for a loan of money or wares without risk to the lender. "Usura est ex mutuo lucrum pacto debitum vel exactum . . . quidquid sorti accedit, subaudi per pactum vel exactionem, usura est, quodcunque nomen sibi imponat." [58]   The emphasis was on *pactum*.   The essence of usury was that it was certain, and that, whether the borrower gained or lost, the usurer took his pound of flesh.   Medieval opinion, which has no objection to rent or profits, provided that they are reasonable—for is not every one in a small way a profit-maker?—has no mercy for the debenture-holder. His crime is that he takes a payment for money which is fixed and certain, and such a payment is usury.

The doctrine was, of course, more complex and more subtle than a bald summary suggests. With the growth of the habit of investment, of a market for capital, and of new forms of economic enterprise such as insurance and exchange business, theory became steadily more elaborate, and schools more sharply divided. The precise meaning and scope of the indulgence extended to the purchase of rent-charges produced one controversy, the foreign exchanges another, the development of *Monts de Piété* a third. Even before the end of the fourteenth century there had been writers who argued that interest was the remuneration of the services rendered by the lender, and who pointed out (though apparently they did not draw the modern corollary) that present are more valuable than future goods.[59] But on the iniquity of payment merely for the act of lending, theological opinion, whether liberal or conservative, was unanimous, and its modern interpreter,[60] who sees in its indulgence to *interesse* the condonation of interest, would have created a scandal in theological circles in any age before that of Calvin. To take usury is contrary to Scripture; it is contrary to Aristotle; it is contrary to nature, for it is to live without labor; it is to sell time, which belongs to God, for the advantage of wicked men; it is to rob those who use the money lent, and to whom, since they make it profitable, the profits should belong; it is unjust in itself, for the benefit of the loan to the borrower cannot exceed the value of the principal sum lent him; it is in defiance of sound juristic principles, for when a loan of money is made, the property in the thing lent passes to the borrower, and why should the creditor demand payment from a man who is merely using what is now his own?

The part played by authority in all this is obvious. There were the texts in Exodus and Leviticus; there was Luke vi. 35—apparently a mistranslation; there was a passage in the *Politics,* which some now say was mistranslated also.[61] But practical considerations contributed more to the doctrine than is sometimes supposed. Its character had been given it in an age in which most loans were not part of a credit system, but an exceptional expedient, and in which it could be said that "he who borrows is always under stress of necessity." If usury were general, it was argued, "men would not give thought to the cultivation of their land, except when they could do nought else, and so there would be so great a famine that all the poor would die of hunger; for even if they could get land to cultivate, they would not be able to get the beasts and implements for cultivating it, since the poor themselves would not have them, and the rich, for the sake both of profit and of security, would put their money into usury rather than into smaller and more risky investments."[62] The man who used these arguments was not an academic dreamer. He was Innocent IV, a consummate man of business, a believer, even to excess, in *Realpolitik,* and one of the ablest statesmen of his day.

True, the Church could not dispense with commercial wickedness in high places. It was too convenient. The distinction between pawnbroking, which is disreputable, and high finance, which is eminently honorable, was as familiar in the Age of Faith as in the twentieth century; and no reasonable judgment of the medieval denunciation of usury is possible, unless it is remembered that whole ranges of financial business escaped from it almost altogether. It was rarely applied to the large-scale transactions of kings, feudal magnates, bishops and abbots. Their subjects, squeezed to pay a foreign money-lender, might grumble or rebel, but, if an Edward III or a Count of Champagne was in the hands of financiers, who could bring either debtor or creditor to book? It was even more rarely applied to the Papacy itself; Popes regularly employed the international banking-houses of the day, with a singular indifference, as was frequently complained, to the morality of their business methods, took them under their special protection, and sometimes enforced the payment of debts by the threat of excommunication. As a rule, in spite of some qualms, the international money-market escaped from it; in the fourteenth century Italy was full of banking-houses doing foreign exchange business in every commercial center from Constantinople to London, and in the great fairs, such as those of Champagne, a special period was regularly set aside for the negotiation of loans and the settlement of debts.[63]

It was not that transactions of this type were expressly excepted; on the contrary, each of them from time to time evoked the protests of moralists. Nor was it mere hypocrisy which caused the traditional doctrine to be repeated by writers who were perfectly well aware that neither commerce nor government could be carried on without credit. It was that the whole body of intellectual assumptions and practical interests, on which the prohibition of usury was based, had reference to a quite different order of economic activities from that represented by loans from great banking-houses to the merchants and potentates who were their clients. Its object was simple and direct—to prevent the well-to-do money-lender from exploiting the necessities of the peasant or the craftsman; its categories, which were quite appropriate to that type of transaction, were those of personal morality. It was in these commonplace dealings among small men that oppression was easiest and its results most pitiable. It was for them that the Church's scheme of economic ethics had been worked out, and with reference to them, though set at naught in high places, it was meant to be enforced, for it was part of Christian charity.

It was enforced partly by secular authorities, partly, in so far as the rivalry of secular authorities would permit it, by the machinery of ecclesiastical discipline. The ecclesiastical legislation on the subject of usury has been so often analyzed that it is needless to do more than allude to it. Early Councils had forbidden usury to be taken by the clergy.[64] The Councils of the twelfth and thirteenth centuries forbid it to be taken by clergy or laity, and lay down rules for dealing with offenders. Clergy who lend money to persons in need, take their possessions in pawn, and receive profits beyond the capital sum lent, are to be deprived of their office.[65] Manifest usurers are not to be admitted to communion or Christian burial; their offerings are not to be accepted; and ecclesiastics who fail to punish them are to be suspended until they make satisfaction to their bishop.[66] The high-water mark of the ecclesiastical attack on usury was probably reached in the legislation of the Councils of Lyons (1274) and of Vienne (1312). The former re-enacted the measures laid down by the third Lateran Council (1175), and supplemented them by rules which virtually made the money-lender an outlaw. No individual or society, under pain of excommunication or interdict, was to let houses to usurers, but was to expel them (had they been admitted) within three months. They were to be refused confession, absolution and Christian burial until they had made restitution, and their wills were to be invalid.[67] The legislation of the Council of Vienne was even more sweeping. Declaring that it has learned with dismay that there are communities which, contrary to human and divine law, sanction usury and compel debtors to observe usurious contracts, it declares that all rulers and magistrates knowingly maintaining such laws are to incur excommunication, and requires the legislation in question to be revoked within three months.  Since the true nature of usurious transactions is often concealed beneath various specious devices, money-lenders are to be compelled by the ecclesiastical authorities to submit their accounts to examination.  Any person obstinately declaring that usury is not a sin is to be punished as a heretic, and inquisitors are to proceed against him *tanquam contra diffamatos vel suspectos de hæresi*.[68]

It would not be easy to find a more drastic example, either of ecclesiastical sovereignty, or of the attempt to assert the superiority of the moral law to economic expediency, than the requirement, under threat of excommunication, that all secular legislation sanctioning usury shall be repealed.  But, for an understanding of the way in which the system was intended to work, the enactments of Councils are perhaps less illuminating than the correspondence between the papal *Curia* and subordinate ecclesiastical authorities on specific cases and questions of interpretation.  Are the heirs of those who have made money by usury bound to make restitution?  Yes, the same penalties are to be applied to them as to the original offenders.  The pious object of ransoming prisoners is not to justify the asking of a price for a loan.  A man is to be accounted a usurer, not only if he charges interest, but if he allows for the element of time in a bargain, by asking a higher price when he sells on credit.  Even when debtors have sworn not to proceed against usurers, the ecclesiastical authorities are to compel the latter to restore their gains, and, if witnesses are terrorized by the protection given to usurers by the powerful, punishment can be imposed without their evidence, provided that the offence is a matter of common notoriety.  An archbishop of Canterbury is reminded that usury is perilous, not only for the clergy, but for all men whatever, and is warned to use ecclesiastical censures to secure the restoration, without the deduction of interest, of property which has been pawned.

Usurers, says a papal letter to the archbishop of Salerno, object to restoring gains, or say that they have not the means; he is to compel all who can to make restitution, either to those from whom interest was taken, or to their heirs; when neither course is possible, they are to give it to the poor; for, as Augustine says, *non remittitur peccatum, nisi restituitur ablatum*. At Genoa, the Pope is informed, a practice obtains of undertaking to pay, at the end of a given term, a higher price for wares than they were worth at the moment when the sale took place. It is not clear that such contracts are necessarily usurious; nevertheless, the sellers run into sin, unless there is a probability that the wares will have changed in value by the time that payment is made; "and therefore your fellow-citizens would show a wise regard for their salvation if they ceased making contracts of the kind, since the thoughts of men cannot be concealed from Almighty God." [69]

It is evident from the number of doubtful cases referred to Rome for decision that the law with regard to usury was not easily administered. It is evident, also, that efforts were made to offer guidance in dealing with difficult and technical problems. In the book of common forms, drawn up in the thirteenth century for the guidance of the papal penitentiary in dealing with hard cases, precedents were inserted to show how usurers should be handled. [70] About the same time appeared St. Raymond's guide to the duties of an archdeacon, which contains a long list of inquiries to be made on visitation, covering every conceivable kind of extortion, and designed to expose the various illusory contracts—fictitious partnerships, loans under the guise of sales, excessive deposits against advances—by which the offence was concealed. [71] Instructions to confessors define in equal detail the procedure to be followed. The confessor, states a series of synodal statutes, is to "make inquiry concerning merchandising, and other things pertaining to avarice and covetousness." Barons and knights are to be requested to state whether they have made ordinances contrary to the liberty of the Church, or refused justice to any man seeking it, or oppressed their subjects with undue tallages, tolls or services. "Concerning burgesses, merchants and officers (*ministrales*) the priest is to make inquiry as to rapine, usury, pledges made by deceit of usury, barratry, false and lying sales, unjust weights and measures, lying, perjury and craft. Concerning cultivators (*agricolas*) he is to inquire as to theft and detention of the property of others, especially with regard to tithes . . . also as to the removing of landmarks and the occupation of other men's land. . . . Concerning avarice it is to be asked in this wise: hast thou been guilty of simony . . . an unjust judge . . . a thief, a robber, a perjurer, a sacrilegious man, a gambler, a remover of landmarks in fields . . . a false merchant, an oppressor of any man and above all of widows, wards and others in misery, for the sake of unjust and greedy gain?" Those guilty of avarice are to do penance by giving large alms, on the principle that "contraries are to be cured with contraries." But there are certain sins for which no true penitence is possible until restitution has been made. Of these usury is one; and usury, it is to be noted, includes, not only what would now be called interest, but the sin of those who, on account of lapse of time, sell dearer and buy cheaper. If for practical reasons restitution is impossible, the offender is to be instructed to require that it shall be made by his heirs, and, when the injured party cannot be found, the money is to be spent, with the advice of the bishop if the sum is large and of the priest if it is small, "on pious works and especially on the poor." [72]

The more popular teaching on the subject is illustrated by the manuals for use in the confessional and by books for the guidance of the devout. The space given in them to the ethics of business was considerable. In the fifteenth

century, Bishop Pecock could meet the Lollards' complaint
that the Scriptures were buried beneath a mass of interpretation, by taking as his illustration the books which had
been written on the text, "Lend, hoping for nothing again,"
and arguing that all this teaching upon usury was little
enough "to answer . . . all the hard, scrupulous doubts
and questions which all day have need to be assoiled in men's
bargains and chafferings together." [73]   A century later there
were regions in which such doctrine was still being rehearsed with all the old rigor.   In 1552 the Parliament which
made the Scottish Reformation was only eight years off.
But the catechism of the archbishop of St. Andrews, which
was drawn up in that year, shows no disposition to compromise with the economic frailties of his fellow-country-
men.   It denounces usurers, masters who withhold wages,
covetous merchants who sell fraudulent wares, covetous
landlords who grind their tenants, and in general—a comprehensive and embarrassing indictment—"all wretches that
will be grown rich incontinent," and all "who may keep
their neighbor from poverty and mischance and do it not." [74]

On the crucial question, how the ecclesiastical courts dealt
in practice with these matters, we have very little light.
They are still almost an unworked field.   On the Continent
we catch glimpses of occasional raids.   Bishops declare war
on notorious usurers, only to evoke reprisals from the
secular authorities, to whom the money-lender is too convenient to be victimized by any one but themselves. [75]   At the
end of the thirteenth century an archbishop of Bourges
makes some thirty-five usurers disgorge at a sitting, [76] and
seventy years later an inquisitor at Florence collects 7,000
florins in two years from usurers and blasphemers. [77]   In
England commercial morality was a debatable land, in which
ecclesiastical and secular authorities contended from time to
time for jurisdiction.   The ecclesiastical courts claimed to
deal with cases of breach of contract in general, on the

ground that they involved *læsio fidei,* and with usury in particular, as an offence against morality specifically forbidden by the canon law. Both claims were contested by the Crown and by municipal bodies. The former, by the Constitutions of Clarendon,[78] had expressly reserved proceedings as to debts for the royal courts, and the same rule was laid down more than once in the course of the next century. The latter again and again forbade burgesses to take proceedings in the courts christian, and fined those who disregarded the prohibition.[79] Both, in spite of repeated protests from the clergy,[80] made good their pretension to handle usurious contracts in secular courts; but neither succeeded in ousting the jurisdiction of the Church. The question at issue was not whether the usurer should be punished —a point as to which there was only one opinion—but who should have the lucrative business of punishing him, and in practice he ran the gauntlet of all and of each. Local authorities, from the City of London to the humblest manorial court, make by-laws against "unlawful chevisance" and present offenders against them.[81] The Commons pray that Lombard brokers may be banished, and that the ordinances of London concerning them may be made of general application.[82] The justices in eyre hear indictments of usurers,[83] and the Court of Chancery handles petitions from victims who can get no redress at common law.[84] And Holy Church, though there seems to be only one example of legislation on the subject by an English Church Council,[85] continues to deal with the usurer after her own manner.

For, in spite of the conflict of jurisdictions, the rising resentment against the ways of ecclesiastical lawyers, and the expanding capitalism of the later Middle Ages, it is evident that commercial cases continued, on occasion at least, to come before the courts christian. Nor, after the middle of the fourteenth century, was their right to try cases of usury contested by the secular authorities. A statute of 1341 enacted that (as laid down long before) the King should have cognizance of usurers dead, and the Church of usurers living. The same reservation of ecclesiastical rights was repeated when the question was taken up a century later under Henry VII, and survived, an antiquated piece of common form, even into the age of lusty capitalism under Elizabeth and James I.[86]

That ecclesiastical authorities had much opportunity of enforcing the canon law in connection with money-lending is improbable. It was naturally in the commercial towns that cases of the kind most frequently arose, and the towns did not look with favor on the interference of churchmen in matters of business. In London, collisions between the courts of the Official, the Mayor and the King were frequent in the early thirteenth century. Men took proceedings before the first, it seems, when a speedy decision was desired, or when their case was of a kind which secular courts were not likely to regard with favor. Thus craftsmen, to give one curious example out of many, were evidently using the courts christian as a means of giving effect to trade union regulations, which were more likely to be punished than enforced by the mayor and aldermen, by the simple device of imposing an oath and proceeding against those who broke it for breach of faith. The smiths, for instance, made a "confederacy," supported by an oath, with the object, as they declared, of putting down night-work, but, as was alleged in court, of preventing any but members of their organization from working at the trade, and summoned blacklegs before the ecclesiastical courts. The spurriers forbade any one to work between sunset and sunrise, and haled an offending journeyman before the archdeacon, with the result that "the said Richard, after being three times warned by the Official, had been expelled from the Church and excommunicated, until he would swear to keep the ordinance." [87]

Even at a later period the glimpses which we catch of the activities of the ecclesiastical jurisdiction are enough to show that it was not wholly a dead letter. Priests accused of usury undergo correction at the hands of their bishops.[88] Petitioners appeal for redress to the Court of Chancery on the ground that they have failed to secure justice in the courts of bishops or archdeacons, where actions on cases of debts or usury have been begun before "spiritual men." [89] The records of ecclesiastical courts show that, though sometimes commercial questions were dismissed as belonging to the secular courts, cases of breach of contract and usury continued, nevertheless, to be settled by them.[90] The disreputable family of Marcroft—William the father was a common usurer, Alice his daughter baked bread at Pentecost, and Edward his son made a shirt on All Saints' Day— is punished by the ecclesiastical court of Whalley as it deserves.[91] At Ripon a usurer and his victim are induced to settle the case out of court.[92] The Commissary of London cites Thomas Hall *super crimine usurariæ pravitatis,* on the ground that, having advanced four shillings on the security of Thomas Foster's belt, he had demanded twelve pence over and above the principal, and suspends him when he does not appear in court.[93] Nor did business of this kind cease with the Reformation. Cases of usury were being heard by ecclesiastical courts under Elizabeth, and even in a great commercial center like the City of London it was still possible in the reign of James I for the Bishop's Commissary to be trying tradesmen for "lending upon pawnes for an excessive gain." [94]

It was not only by legal penalties, however, that an attempt was made to raise a defensive barrier against the exactions of the money-lender. From a very early date there was a school of opinion which held that, in view of the various stratagems by which usurious contracts could be "colored," direct prohibition was almost necessarily impotent, and which favored the policy of providing facilities for borrowing on more reasonable terms than could be obtained from the money-lender.  Ecclesiastics try, in fact, to turn the flank of the usurer by establishing institutions where the poor can raise capital cheaply.  Parishes, religious fraternities, gilds, hospitals and perhaps monasteries lend corn, cattle and money.[95]  In England, bishops are organizing such loans with papal approval in the middle of the thirteenth century,[96] and two centuries later, about 1462, the Franciscans lead the movement for the creation of *Monts de Piété,* which, starting in Italy, spread by the first half of the sixteenth century to France, Germany, and the Low Countries, and, though never taken up in England—for the Reformation intervened—supplied a topic of frequent comment and eulogy to English writers on economic ethics.[97]  The canon law on the subject of money-lending underwent a steady development, caused by the necessity of adapting it to the increasing complexity of business organization, down at least to the Lateran Council of 1515.  The ingenuity with which professional opinion elaborated the code was itself a proof that considerable business—and fees—were the result of it, for lawyers do not serve God for naught.  The canonists, who had a bad reputation with the laity, were not, to put it mildly, more innocent than other lawyers in the gentle art of making business.  The Italians, in particular, as was natural in the financial capital of Europe, made the pace, and Italian canonists performed prodigies of legal ingenuity.  In England, on the other hand, either because Englishmen were unusually virtuous, or, as a foreigner unkindly said, because "they do not fear to make contracts on usury," [98] or, most probably, because English business was a conservative and slow-going affair, the English canonist Lyndwood is content to quote a sentence from an English archbishop of the thirteenth century and to leave it at that.[99]

But, however lawyers might distinguish and refine, the essential facts were simple. The Church sees buying and selling, lending and borrowing, as a simple case of neighborly or unneighborly conduct. Though a rationalist like Bishop Pecock may insist that the rich, as such, are not hateful to God,[100] it has a traditional prejudice against the arts by which men—or at least laymen—acquire riches, and is apt to lump them together under the ugly name of avarice. Merchants who organize a ring, or money-lenders who grind the poor, it regards, not as business strategists, but as *nefandæ belluæ*—monsters of iniquity. As for grocers and victualers "who conspire wickedly together that none shall sell better cheap than another," and speculators "who buy up corn, meat and wine . . . to amass money at the cost of others," they are "according to the laws of the Church no better than common criminals."[101] So, when the price of bread rises, or when the London fruiterers, persuaded by one bold spirit that they are "all poor and caitiffs on account of their own simplicity, and if they would act on his advice they would be rich and powerful,"[102] form a combine, to the great loss and hardship of the people, burgesses and peasants do not console themselves with the larger hope that the laws of supply and demand may bring prices down again. Strong in the approval of all good Christians, they stand the miller in the pillory, and reason with the fruiterers in the court of the mayor. And the parish priest delivers a sermon on the sixth commandment, choosing as his text the words of the Book of Proverbs, "Give me neither riches nor poverty, but enough for my sustenance."

### III. THE IDEAL AND THE REALITY

Such, in brief outline, was the background of economic thought which the sixteenth century inherited, and which it brought to the bewildering changes in land tenure, in prices, in commercial and financial organization, that made the age a watershed in economic development. It is evident that the whole implication of this philosophy was, on one side, intensely conservative. There was no question of progress, still less of any radical social reconstruction. In the numerous heretical movements of the Middle Ages social aspirations were often combined with criticisms of the luxury and pomp of the ecclesiastical hierarchy. The official Church, to which independence of thought among the lower orders was but little less abhorrent when it related to their temporal well-being than when it was concerned with their eternal salvation, frowned upon these dangerous speculations, and sometimes crushed them with a ferocity as relentless as the most savage of the White Terrors of modern history has shown to the most formidable of insurrections.

Intellectually, religious opinion endorsed to the full the static view, which regarded the social order as a thing unalterable, to be accepted, not to be improved. Except on rare occasions, its spokesmen repeated the conventional doctrine, according to which the feet were born to labor, the hands to fight, and the head to rule. Naturally, therefore, they denounced agitations, like the communal movement,[103] designed to overturn that natural order, though the rise of the Free Cities was one of the glories of medieval Europe and the germ of almost every subsequent advance in civilization. They referred to questions of economic conduct, not because they were anxious to promote reforms, but because they were concerned with the maintenance of traditional standards of personal morality, of which economic conduct formed an important part.

Practically, the Church was an immense vested interest, implicated to the hilt in the economic fabric, especially on the side of agriculture and land tenure. Itself the greatest of landowners, it could no more quarrel with the feudal structure than the Ecclesiastical Commission, the largest of mineral owners today, can lead a crusade against royalties. The persecution of the Spiritual Franciscans, who dared, in defiance of the bull of John XXII, to maintain St. Francis' rule as to evangelical poverty, suggests that doctrines impugning the sanctity of wealth resembled too closely the teaching of Christ to be acceptable to the princes of the Christian Church.

The basis of the whole medieval economic system, under which, except in Italy and Flanders, more than nine-tenths of the population consisted of agriculturists, had been serfdom or villeinage. Confronted in the sixteenth century with the unfamiliar evils of competitive agriculture, conservative reformers were to sigh for the social harmonies of a vanished age, which "knyt suche a knott of colaterall amytie betwene the Lordes and the tenaunts that the Lorde tendered his tenaunt as his childe, and the tenaunts againe loved and obeyed the Lorde as naturellye as the childe the father." [104] Their idealization of the past is as misleading, as an account of the conditions of previous centuries, as it is illuminating as a comment upon those of their own. In reality, so far as the servile tenants, who formed the bulk of medieval agriculturists, were concerned, the golden age of peasant prosperity is, except here and there, a romantic myth, at which no one would have been more surprised than the peasants themselves. The very essence of feudal property was exploitation in its most naked and shameless form, compulsory labor, additional *corvées* at the very moments when the peasant's labor was most urgently needed on his own holding, innumerable dues and payments, the obligation to grind at the lord's mill and bake at the lord's oven, the private justice of the lord's court. The custom of the manor, the scarcity of labor, and, in England, the steadily advancing encroachments of the royal courts, blunted the edge of the system, and in fifteenth-century England a prosperous yeomanry was rising on its ruins. But, during the greater part of the Middle Ages, its cumulative weight had been, nevertheless, immense. Those who lived under it had no illusions as to its harshness. The first step which the peasant who had saved a little money took was to buy himself out of the obligation to work on the lord's demesne. The Peasants' Revolt in England, the *Jacquerie* in France and the repeated risings of the German peasantry reveal a state of social exasperation which has been surpassed in bitterness by few subsequent movements.

It is natural to ask (though some writers on medieval economics refrain from asking) what the attitude of religious opinion was towards serfdom. And it is hardly possible to answer that question except by saying that, apart from a few exceptional individuals, religious opinion ignored it. True, the Church condemned arbitrary tallages, and urged that the serf should be treated with humanity. True, it described the manumission of serfs as an act of piety, like gifts to the poor. For serfs are not "living tools," but men; in the eyes of God all men are serfs together, *conservi,* and in the Kingdom of Heaven Lazarus is before Dives.[105] True, villeinage was a legal, not an economic, category; in the England of the fourteenth century there were serfs who were rich men. But to release the individual is not to condemn the institution. Whatever "mad priests" might say and do, the official Church, whose wealth consisted largely of villeins, walked with circumspection.

The canon law appears to have recognized and enforced serfdom.[106] Few prominent ecclesiastics made any pronouncement against it. Aquinas explains it as the result of sin, but that does not prevent his justifying it on economic grounds.[107] Almost all medieval writers appear to assume it or excuse it.   Ecclesiastical landlords, though perhaps somewhat more conservative in their methods, seem as a whole to have been neither better nor worse than other landlords.   *Rustica gens optima flens, pessima gaudens,* was a sentiment which sometimes appealed, it is to be feared, to the children of light concerned with rent rolls and farming profits, not less than to the feudal aristocracy, with whom the heads of the ecclesiastical hierarchy were inextricably intermingled.   When their chance came, John Nameless, and John the Miller, and John Carter, who may be presumed to have known their friends, burned the court rolls of an abbot of St. Albans, and cut off the head of an archbishop, and ran riot on the estates of an abbot of Kempten, with not less enthusiasm than they showed in plundering their lay exploiters.   It was not the Church, but revolting peasants in Germany and England, who appealed to the fact that "Christ has made all men free"; [108] and in Germany, at least, their ecclesiastical masters showed small mercy to them.   The disappearance of serfdom— and, after all, it did not disappear from France till late in the eighteenth century, and from Germany till the nineteenth—was part of a general economic movement, with which the Church had little to do, and which churchmen, as property-owners, had sometimes resisted.   It owed less to Christianity than to the humanitarian liberalism of the French Revolution.

The truth was that the very triumph of the Church closed its mouth.   The Church of the third century, a minority of believers confronted with an alien civilization, might protest and criticize.   But, when the whole leaven was mixed with the lump, when the Church was regarded, not as *a* society, but as society itself, it was inevitably diluted by the mass which it absorbed.   The result was a compromise—a compromise of which the critic can say, "How much that was intolerable was accepted!" and the eulogist, "How much that was intolerable was softened!"

Both critic and eulogist are right. For if religious opinion acquiesced in much, it also claimed much, and the habit of mind which made the medieval Church almost impotent when dealing with the serried abuses of the medieval land system was precisely that which made it strong, at least in theory, in dealing with the economic transactions of the individual. In the earlier Middle Ages it had stood for the protection of peaceful labor, for the care of the poor, the unfortunate and the oppressed—for the ideal, at least, of social solidarity against the naked force of violence and oppression. With the growing complexity of economic civilization, it was confronted with problems not easily handled by its traditional categories. But, if applied capriciously, they were not renounced, and the world of economic morality, which baffles us today, was in its turn converted by it into a new, though embarrassing, opportunity. Whatever emphasis may be laid—and emphasis can hardly be too strong—upon the gulf between theory and practice, the qualifications stultifying principles, and the casuistry by which the work of canonists, not less than of other lawyers, was disfigured, the endeavor to draw the most commonplace of human activities and the least tractable of human appetites within the all-embracing circle of a universal system still glows through it all with a certain tarnished splendor. When the distinction between that which is permissible in private life and that which is permissible in business offers so plausible an escape from the judgment pronounced on covetousness, it is something to have insisted that the law of charity is binding on the second not less than on the first. When the austerity of principles can be evaded by treating them as applicable only to those relations of life in which their application is least exacting, it is something to have attempted to construct a system tough enough to stand against commercial unscrupulousness, but yet sufficiently elastic to admit any legitimate transaction.  If it is proper to insist on the prevalence of avarice and greed in high places, it is not less important to observe that men called these vices by their right names, and had not learned to persuade themselves that greed was enterprise and avarice economy.

Such antitheses are tempting, and it is not surprising that some writers should have dwelt upon them.  To a generation disillusioned with free competition, and disposed to demand some criterion of social expediency more cogent than the verdict of the market, the jealous and cynical suspicion of economic egotism, which was the prevalent mood of the Middle Ages, is more intelligible than it was to the sanguine optimists of the Age of Reason, which, as far as its theory of the conduct of men in society is concerned, deserves much more than the thirteenth century to be described as the Age of Faith.  In the twentieth century, with its trusts and combines, its control of industry by business and of both by finance, its attempts to fix fair wages and fair prices, its rationing and food controls and textile controls, the economic harmonies are, perhaps, a little blown upon.  The temper in which it approaches questions of economic organization appears to have more affinity with the rage of the medieval burgess at the uncharitable covetousness of the usurer and the engrosser, than it has with the confidence reposed by its innocent grandfathers in the infallible operations of the invisible hand.

The resemblance, however, though genuine, is superficial, and to over-emphasize it is to do less than justice to precisely those elements in medieval thought which were most characteristic.  The significance of its contribution consists, not in its particular theories as to prices and interest, which recur in all ages, whenever the circumstances of the economic environment expose consumer and borrower to extortion, but in its insistence that society is a spiritual organism, not an economic machine, and that economic activity, which is one subordinate element within a vast and complex unity, requires to be controlled and repressed by reference to the moral ends for which it supplies the material means. So merciless is the tyranny of economic appetites, so prone to self-aggrandizement the empire of economic interests, that a doctrine which confines them to their proper sphere, as the servant, not the master, of civilization, may reasonably be regarded as among the pregnant truisms which are a permanent element in any sane philosophy. Nor is it, perhaps, as clear today as it seemed a century ago, that it has been an unmixed gain to substitute the criterion of economic expediency, so easily interpreted in terms of quantity and mass, for the conception of a rule of life superior to individual desires and temporary exigencies, which was what the medieval theorist meant by "natural law."

When all is said, the fact remains that, on the small scale involved, the problem of moralizing economic life was faced and not abandoned. The experiment may have been impracticable, and almost from the first it was discredited by the notorious corruption of ecclesiastical authorities, who preached renunciation and gave a lesson in greed. But it had in it something of the heroic, and to ignore the nobility of the conception is not less absurd than to idealize its practical results. The best proof of the appeal which the attempt to subordinate economic interests to religion had made is the persistence of the same attempt among reformers, to whom the Pope was anti-Christ and the canon law an abomination and the horror of decent men when, in the sixteenth century, its breakdown became too obvious to be contested.

# CHAPTER II

## THE CONTINENTAL REFORMERS

"Neither the Church of Christ, nor a Christian Commonwealth, ought to tolerate such as prefer private gain to the public weal, or seek it to the hurt of their neighbours."

BUCER, *De Regno Christi.*

# CHAPTER II

Lord Acton, in an unforgettable passage in his *Inaugural Lecture on the Study of History,* has said that "after many ages persuaded of the headlong decline and impending dissolution of society, and governed by usage and the will of masters who were in their graves, the sixteenth century went forth armed for untried experience, and ready to watch with hopefulness a prospect of incalculable change." [1] His reference was to the new world revealed by learning, by science, and by discovery. But his words offer an appropriate text for a discussion of the change in the conception of the relations between religion and secular interests which took place in the same period. Its inevitable consequence was the emergence, after a prolonged moral and intellectual conflict, of new conceptions of social expediency and of new lines of economic thought.

The strands in this movement were complex, and the formula which associates the Reformation with the rise of economic individualism is no complete explanation. Systems prepare their own overthrow by a preliminary process of petrifaction. The traditional social philosophy was static, in the sense that it assumed a body of class relations sharply defined by custom and law, and little affected by the ebb and flow of economic movements. Its weakness in the face of novel forces was as obvious as the strain put upon it by the revolt against the source of ecclesiastical jurisprudence, the partial discredit of the canon law and of ecclesiastical discipline, and the rise of a political science equipped from the arsenals of antiquity. But it is not to under-estimate the effect of the Reformation to say that the principal causes making the age a watershed, from which new streams of social theory descend, lay in another region.  Mankind does not reflect upon questions of economic and social organization until compelled to do so by the sharp pressure of some practical emergency.  The sixteenth century was an age of social speculation for the same reason as the early nineteenth —because it was an age of social dislocation.  The retort of conservative religious teachers to a spirit which seems to them the triumph of Mammon produces the last great literary expression of the appeal to the average conscience which had been made by an older social order.  The practical implications of the social theory of the Middle Ages are stated more clearly in the sixteenth century than even in its zenith, because they are stated with the emphasis of a creed which is menaced.

## I.  THE ECONOMIC REVOLUTION

The religious revolution of the age came on a world heaving with the vastest economic crisis that Europe had experienced since the fall of Rome.  Art and scientific curiosity and technical skill, learning and statesmanship, the scholarship which explored the past and the prophetic vision which pierced the future, had all poured their treasures into the sumptuous shrine of the new civilization.  Behind the genii of beauty and wisdom who were its architects there moved a murky, but indispensable, figure.  It was the demon whom Dante had met muttering gibberish in the fourth circle of the Inferno, and whom Sir Guyon was to encounter three centuries later, tanned with smoke and seared with fire, in a cave adjoining the mouth of hell.  His uncouth labors quarried the stones which Michael Angelo was to raise, and sank deep in the Roman clay the foundations of the walls to be adorned by Raphael.

For it was the mastery of man over his environment which heralded the dawn of the new age, and it was in the stress of expanding economic energies that this mastery was proved and won. Like sovereignty in a feudal society, the economic efforts of the Middle Ages, except in a few favored spots, had been fragmentary and decentralized. Now the scattered raiders were to be organized and disciplined; the dispersed and irregular skirmishes were to be merged in a grand struggle, on a front which stretched from the Baltic to the Ganges and from the Spice Islands to Peru. Every year brought the news of fresh triumphs. The general who marshaled the host and launched the attack was economic power.

Economic power, long at home in Italy, was leaking through a thousand creeks and inlets into western Europe, for a century before, with the climax of the great Discoveries, the flood came on breast-high. Whatever its truth as a judgment on the politics of the fifteenth century, the conventional verdict on its futility does scanty justice to its economic significance. It was in an age of political anarchy that the forces destined to dominate the future tried their wings. The era of Columbus and Da Gama was prepared by the patient labor of Italian cartographers and Portuguese seamen, as certainly as was that of Crompton and Watt by the obscure experiments of nameless predecessors.

The master who set the problem that the heroes of the age were to solve was material necessity. The Europe of the earlier Middle Ages, like the world of the twentieth century, had been a closed circle. But it had been closed, not by the growth of knowledge, but by the continuance of ignorance; and, while the latter, having drawn the whole globe into a single economic system, has no space left for fresh expansion, for the former, with the Mediterranean as its immemorial pivot, expansion had hardly begun. Tapping the wealth of the East by way of the narrow apertures in the

Levant, it resembled, in the rigidity of the limits imposed on its commercial strategy, a giant fed through the chinks of a wall.

As was the general scheme, so were the details; inelastic in its external, Europe was hardly more flexible in its internal, relations. Its primary unit had been the village; and the village, a community of agrarian shareholders fortified by custom, had repressed with a fury of virtuous unanimity the disorderly appetites which menaced its traditional routine with the evil whose name is Change. Beyond the village lay the greater, more privileged, village called the borough, and the brethren of borough and gild had turned on the foreign devil from upland and valley a face of flint. Above both were the slowly waking nations. Nationalism was an economic force before nationality was a political fact, and it was a sound reason for harrying a competitor that he was a Florentine or a man of the Emperor. The privileged colony with its depôt, the Steel-yard of the Hanseatic League, the Fondaco Tedesco of the south Germans, the Factory of the English Merchant Adventurers, were but tiny breaches in a wall of economic exclusiveness. Trade, as in modern Turkey or China, was carried on under capitulations.

This narrow framework had been a home. In the fifteenth century it was felt to be a prison. Expanding energies pressed against the walls; restless appetites gnawed and fretted wherever a crack in the surface offered room for erosion. Long before the southward march of the Turks cut the last of the great routes from the East, the Venetian monopoly was felt to be intolerable. Long before the plunder of Mexico and the silver of Potosi flooded Europe with treasure, the mines of Germany and the Tyrol were yielding increasing, if still slender, streams of bullion, which stimulated rather than allayed its thirst.[2] It was not the lords of great estates, but eager and prosperous peasants, who in England first nibbled at commons and undermined the manorial custom, behind which, as behind a dyke, their small savings had been accumulated. It was not great capitalists, but enterprising gildsmen, who began to make the control of the fraternity the basis of a system of plutocratic exploitation, or who fled, precocious individualists, from the fellowship of borough and craft, that they might grow to what stature they pleased in rural isolation. It was not even the Discoveries which first began the enormous tilt of economic power from south and east to north and west. The records of German and English trade suggest that the powers of northern Europe had for a century before the Discoveries been growing in wealth and civilization,[3] and for a century after them English economic development was to be as closely wedded to its continental connections as though Diaz had never rounded the Cape, nor Columbus praised Heaven for leading him to the shores of Zayton and Guinsay. First attempted as a counterpoise to the Italian monopolist, then pressed home with ever greater eagerness to turn the flank of the Turk, as his strangle-hold on the eastern commerce tightened, the Discoveries were neither a happy accident nor the fruit of the disinterested curiosity of science. They were the climax of almost a century of patient economic effort. They were as practical in their motive as the steam-engine.

The result was not the less sensational because it had been long prepared. Heralded by an economic revolution not less profound than that of three centuries later, the new world of the sixteenth century took its character from the outburst of economic energy in which it had been born. Like the nineteenth century, it saw a swift increase in wealth and an impressive expansion of trade, a concentration of financial power on a scale unknown before, the rise, amid fierce social convulsions, of new classes and the depression of old, the triumph of a new culture and system of ideas amid struggles not less bitter.

It was an age of economic, not less than of political, sensations, which were recorded in the letter-books [4] of business men as well as in the state papers of Governments. The decline of Venice and of the south German cities which had distributed the products that Venice imported, and which henceforward must either be marooned far from the new trade routes or break out to the sea, as some of them did, by way of the Low Countries; the new economic imperialism of Portugal and Spain; the outburst of capitalist enterprise in mining and textiles; the rise of commercial companies, no longer local but international, and based, not merely on exclusive privileges, but on the power of massed capital to drive from the field all feebler competitors; a revolution in prices which shattered all customary relationships; the collapse of medieval rural society in a nightmare of peasants' wars; the subjection of the collegiate industrial organization of the Middle Ages to a new money-power; the triumph of the State and its conquest, in great parts of Europe, of the Church—all were crowded into less than two generations. A man who was born when the Council of Basel was sitting saw also, if he lived to a ripe old age, the dissolution of the English monasteries. At the first date Portuguese explorers had hardly passed Sierra Leone; at the second Portugal had been the master of an Indian Empire for almost a generation. In the intervening three-quarters of a century the whole framework of European civilization had been transformed.

Compared with the currents which raced in Italy, or Germany, or the Low Countries, English life was an economic back-water. But even its stagnant shallows were stirred by the eddy and rush of the continental whirlpool. When Henry VII came to the throne, the economic organization of the country differed but little from that of the age of Wyclif. When Henry VIII died, full of years and sin, some of the main characteristics which were to distinguish it till the advent of steam-power and machinery could already, though faintly, be descried. The door that remained to be unlocked was colonial expansion, and forty years later the first experiments in colonial expansion had begun.

The phenomenon which dazzled contemporaries was the swift start into apparent opulence, first of Portugal and then of Spain. The nemesis of parasitic wealth was not discerned, and it was left for the cynical rationalism of an ambassador of that commercial republic, in comparison with whose hoary wisdom the new plutocrats of the West were meddlesome children, to observe that the true mines of the Spanish Empire lay, not in America, but in the sodden clay of the water-logged Netherlands.[5] The justice of the criticism was revealed when Spain, a corpse bound on the back of the most liberal and progressive community of the age, completed her own ruin by sacking the treasury from which, far more than from Potosi, her wealth had been drawn. But the beginnings of that long agony, in which the power-house of European enterprise was to be struck with paralysis, lay still in the future, and later generations of Spaniards looked back with pardonable exaggeration on the closing years of Charles V as a golden age of economic prosperity. Europe as a whole, however lacerated by political and religious struggles, seemed to have solved the most pressing of the economic problems which had haunted her in the later Middle Ages. During a thousand years of unresting struggle with marsh and forest and moor she had colonized her own waste places. That tremendous achievement almost accomplished, she now turned to the task of colonizing the world. No longer on the defensive, she entered on a phase of economic expansion which was to grow for the next four hundred years, and which only in the twentieth century was to show signs of drawing towards its close. Once a year she was irrigated with the bullion of America, once a year she was enriched with a golden harvest from the East. The period of mere experiment over, and the new connections firmly established, she appeared to be in sight of an economic stability based on broader foundations than ever before.

Portugal and Spain held the keys of the treasure-house of East and West. But it was not Portugal, with her tiny population, and her empire that was little more than a line of forts and factories 10,000 miles long, nor Spain, for centuries an army on the march, and now staggering beneath the responsibilities of her vast and scattered empire, devout to fanaticism, and with an incapacity for economic affairs which seemed almost inspired, who reaped the material harvest of the empires into which they had stepped, the one by patient toil, the other by luck. Gathering spoils which they could not retain, and amassing wealth which slipped through their fingers, they were little more than the political agents of minds more astute and characters better versed in the arts of peace. Every period and society has some particular center, or institution, or social class, in which the characteristic qualities of its genius seem to be fixed and embodied. In the Europe of the early Renaissance the heart of the movement had been Italy. In the Europe of the Reformation it was the Low Countries. The economic capital of the new civilization was Antwerp. The institution which best symbolized its eager economic energies was the international money-market and produce-exchange. Its typical figure, the paymaster of princes, was the international financier.

Before it was poisoned by persecution, revolution and war, the spirit of the Netherlands found its purest incarnation in Erasmus, a prophet without sackcloth and a reformer untouched by heat or fury, to the universal internationalism of whose crystal spirit the boundaries of States were a pattern scrawled to amuse the childish malice of princes. Of that cosmopolitan country, destined to be the refuge of the international idea when outlawed by every other power in Europe, Antwerp, "a home common to all nations," was the most cosmopolitan city. Made famous as a center of learning by Plantin's press, the metropolis of painting in a country where painting was almost a national industry, it was at once the shrine to which masters like Cranach, Dürer and Holbein made their pilgrimage of devotion, and an asylum which offered to the refugees of less happy countries a haven as yet undisturbed by any systematic campaign to stamp out heresy. In the exuberance of its intellectual life, as in the glitter of its material prosperity, the thinker and the reformer found a spiritual home, where the energies of the new age seemed gathered for a bound into that land of happiness and dreams, for the scene of which More, who knew his Europe, chose as the least incredible setting the garden of his lodgings at Antwerp.

The economic preëminence of Antwerp owed much to the industrial region behind it, from which the woollen and worsteds of Valenciennes and Tournai, the tapestries of Brussels and Oudenarde, the iron of Namur, and the munitions of the Black Country round Liége, poured in an unceasing stream on to its quays.[6] But Antwerp was a European, rather than a Flemish, metropolis. Long the competitor of Bruges for the reception of the two great currents of trade from the Mediterranean and the Baltic, which met in the Low Countries, by the last quarter of the fifteenth century she had crushed her rival. The Hanse League maintained a depôt at Antwerp; Italian banking firms in increasing numbers opened businesses there; the English Merchant Adventurers made it the entrepôt through which English cloth, long its principal import, was distributed to northern Europe; the copper market moved from Venice to Antwerp in the nineties. Then came the great Discoveries, and Antwerp, the first city to tap the wealth, not of an inland sea, but of the ocean, stepped into a position of unchallenged preëminence almost unique in European history. The long sea-roads which ran east and west met and ended in its harbors. The Portuguese Government made it in 1503 the depôt of the Eastern spice trade. From the accession of Charles V it was the commercial capital of the Spanish Empire, and, in spite of protests that the precious metals were leaving Spain, the market for American silver. Commerce, with its demand for cheap and easy credit, brought finance in its train. The commercial companies and banking houses of south Germany turned from the dwindling trade across the Alps, to make Antwerp the base for financial operations of unexampled magnitude and complexity.[7]

In such an economic forcing-house new philosophies of society, like new religious creeds, found a congenial soil. Professor Pirenne has contrasted the outlook of the medieval middle class, intent on the conservation of corporate and local privileges, with that of the new plutocracy of the sixteenth century, with its international ramifications, its independence of merely local interests, its triumphant vindication of the power of the capitalist to dispense with the artificial protection of gild and borough and carve his own career.[8] "No one can deny," wrote the foreign merchants at Antwerp to Philip II, in protest against an attempt to interfere with the liberty of exchange transactions, "that the cause of the prosperity of this city is the freedom granted to those who trade there."[9] Swept into wealth on the crest of a wave of swiftly expanding enterprise, which a century before would have seemed the wildest of fantasies, the liberal *bourgeoisie* of Antwerp pursued, in the teeth of all precedents, a policy of practical individualism, which would have been met in any other city by rebellion, making terms with the levelling encroachments of the Burgundian monarchy, which were fought by their more conservative neigh-

bors, lowering tariffs and extinguishing private tolls, wel-
coming the technical improvements which elsewhere were
resisted, taming the turbulent independence of the gilds, and
throwing open to alien and citizen alike the new Exchange,
with its significant dedication: *Ad usum mercatorum
cuiusque gentis ac linguae.*

For, if Antwerp was the microcosm which reflected the
soul of commercial Europe, the heart of Antwerp was its
Bourse. The causes which made financial capitalism as
characteristic of the age of the Renaissance, as industrial
capitalism was to be of the nineteenth century, consisted
partly in the mere expansion in the scale of commercial enterprise. A steady flow of capital was needed to finance the
movement of the produce handled on the world-market, such
as the eastern spice crop—above all pepper, which the impecunious Portuguese Government sold in bulk, while it
was still on the water, to German syndicates—copper, alum,
the precious metals, and the cloth shipped by the English
Merchant Adventurers. The cheapening of bullion and the
rise in prices swelled the profits seeking investment; the
growth of an international banking system mobilized immense resources at the strategic points; and, since Antwerp
was the capital of the European money-market, the bill on
Antwerp was the commonest form of international currency. Linked together by the presence in each of the
great financial houses of the Continent, with liquid funds
pouring in from mines in Hungary and the Tyrol, trading
ventures in the East, taxes wrung from Spanish peasants,
speculations on the part of financiers, and savings invested
by the general public, Antwerp, Lyons, Frankfurt and
Venice, and, in the second rank, Rouen, Paris, Strassburg,
Seville and London, had developed by the middle of the
century a considerable class of financial specialists, and a
financial technique, identical, in all essentials, with that of
the present day. They formed together the departments of

an international clearing-house, where bills could be readily discounted, drafts on any important city could be obtained, and the paper of merchants of almost every nationality changed hands.[10]

Nourished by the growth of peaceful commerce, the financial capitalism of the age fared not less sumptuously, if more dangerously, at the courts of princes. Mankind, it seems, hates nothing so much as its own prosperity. Menaced with an accession of riches which would lighten its toil, it makes haste to redouble its labors, and to pour away the perilous stuff, which might deprive of plausibility the complaint that it is poor. Applied to the arts of peace, the new resources commanded by Europe during the first half of the sixteenth century might have done something to exorcise the specters of pestilence and famine, and to raise the material fabric of civilization to undreamed-of heights. Its rulers, secular and ecclesiastical alike, thought otherwise. When pestilence and famine were ceasing to be necessities imposed by nature, they reëstablished them by political art.

The sluice which they opened to drain away each new accession of superfluous wealth was war. "Of all birds," wrote the sharpest pen of the age, "the eagle alone has seemed to wise men the type of royalty—not beautiful, not musical, not fit for food, but carnivorous, greedy, hateful to all, the curse of all, and, with its great powers of doing harm, surpassing them in its desire of doing it." [11] The words of Erasmus, uttered in 1517, were only too prophetic. For approximately three-quarters both of the sixteenth and of the seventeenth centuries, Europe tore itself to pieces. In the course of the conflict the spiritual fires of Renaissance and Reformation alike were trampled out beneath the feet of bravos as malicious and mischievous as the vain, bloody-minded and futile generals who strut and posture, to the hateful laughter of Thersites, in the most despairing of Shakespeare's tragedies. By the middle of the sixteenth century the English Government, after an orgy of debasement and confiscation, was in a state of financial collapse, and by the end of it Spain, the southern Netherlands including Antwerp, and a great part of France, including the financial capital of southern Europe, Lyons, were ruined. By the middle of the seventeenth century wide tracts of Germany were a desert, and by the end of it the French finances had relapsed into worse confusion than that from which they had been temporarily rescued by the genius of Colbert. The victors compared their position with that of the vanquished, and congratulated themselves on their spoils. It rarely occurred to them to ask what it would have been, had there been neither victors nor vanquished, but only peace.

It is possible that the bankruptcies of Governments have, on the whole, done less harm to mankind than their ability to raise loans, and the mobilization of economic power on a scale unknown before armed the fierce nationalism of the age with a weapon more deadly than gunpowder and cannon. The centralized States which were rising in the age of the Renaissance were everywhere faced with a desperate financial situation. It sprang from the combination of modern administrative and military methods with medieval systems of finance. They entrusted to bureaucracies work which, if done at all, had formerly been done as an incident of tenure, or by boroughs and gilds; officials had to be paid. They were constantly at war; and the new technique of war, involving the use of masses of professional infantry and artillery—which Rabelais said was invented by the inspiration of the devil, as a counterpoise to the invention of printing inspired by God—was making it, as after 1870, a highly capitalized industry. Government after Government, undeterred, with rare exceptions, by the disasters of its neighbors, trod a familiar round of expedients, each of which was more disastrous than the last. They hoarded treasure, only to see the accumulations of a thrifty Henry VII or Frederick III dissipated by a Henry VIII or a Maximilian. They debased the currency and ruined trade. They sold offices, or established monopolies, and crushed the tax-payer beneath a load of indirect taxation. They plundered the Church, and spent gorgeously as income property which should have been treated as capital. They parted with Crown estates, and left an insoluble problem to their successors.

These agreeable devices had, however, obvious limits. What remained, when they were exhausted, was the money-market, and to the rulers of the money-market sooner or later all States came. Their dependence on the financier was that of an Ismail or an Abdul, and its results were not less disastrous. Naturally, the City interest was one of the great Powers of Europe. Publicists might write that the new Messiah was the Prince, and reformers that the Prince was Pope. But behind Prince and Pope alike, financing impartially Henry VIII, Edward VI and Elizabeth, Francis, Charles and Philip, stood in the last resort a little German banker, with branches in every capital in Europe, who played in the world of finance the part of the *condottieri* in war, and represented in the economic sphere the morality typified in that of politics by Machiavelli's Prince. Compared with these financial dynasties, Hapsburgs, Valois and Tudors were puppets dancing on wires held by a money-power to which political struggles were irrelevant except as an opportunity for gain.

The financier received his payment partly in cash, partly in concessions, which still further elaborated the network of financial connections that were making Europe an economic unity. The range of interests in which the German banking houses were involved is astonishing. The Welsers had invested in the Portuguese voyage of 1505 to the East Indies, financed an expedition, half commercial, half military, to Venezuela in 1527, were engaged in the spice trade between Lisbon, Antwerp and south Germany, were partners in silver and copper mines in the Tyrol and Hungary, and had establishments, not only at Lisbon and Antwerp, but in the principal cities of Germany, Italy and Switzerland. The careers of the Hochstetters, Haugs, Meutings and Imhofs were much the same. The Fuggers, thanks to judicious loans to Maximilian, had acquired enormous concessions of mineral property, farmed a large part of the receipts drawn by the Spanish Crown from its estates, held silver and quicksilver mines in Spain, and controlled banking and commercial businesses in Italy, and, above all, at Antwerp. They advanced the money which made Albrecht of Brandenburg archbishop of Mainz; repaid themselves by sending their agent to accompany Tetzel on his campaign to raise money by indulgences and taking half the proceeds; provided the funds with which Charles V bought the imperial crown, after an election conducted with the publicity of an auction and the morals of a gambling hell; browbeat him, when the debt was not paid, in the tone of a pawnbroker rating a necessitous client; and found the money with which Charles raised troops to fight the Protestants in 1552. The head of the firm built a church and endowed an almshouse for the aged poor in his native town of Augsburg. He died in the odor of sanctity, a good Catholic and a Count of the Empire, having seen his firm pay 54 per cent. for the preceding sixteen years.[12]

## II.  LUTHER

Like the rise of the great industry three centuries later, the economic revolution which accompanied the Renaissance gave a powerful stimulus to speculation. Both in Germany and in England, the Humanists turned a stream of

pungent criticism on the social evils of their age. Mercantilist thinkers resharpened an old economic weapon for the
armory of princes. Objective economic analysis, still in its
infancy, received a new impetus from the controversies of
practical men on the rise in prices, on currency, and on the
foreign exchanges.

The question of the attitude which religious opinion
would assume towards these new forces was momentous.
It might hail the outburst of economic enterprise as an instrument of wealth and luxury, like the Popes who revelled
in the rediscovery of classical culture. It might denounce
it as a relapse into a pagan immorality, like the Fathers
who had turned with a shudder from the material triumphs
of Rome. It might attempt to harness the expanding energies to its own conception of man's spiritual end, like the
Schoolmen who had stretched old formulæ to cover the new
forces of capital and commerce. It could hardly ignore
them. For, in spite of Machiavelli, social theory was only
beginning to emancipate itself from the stiff ecclesiastical
framework of the Middle Ages. The most systematic
treatment of economic questions was still that contained in
the work of canonists, and divines continued to pronounce
judgment on problems of property and contract with the
same assurance as on problems of theology.

Laymen might dispute the content of their teaching and
defy its conclusions. But it was rarely, as yet, that they
attacked the assumption that questions of economic conduct belonged to the province of the ecclesiastical jurist.
Bellarmin complained with some asperity of the intolerable
complexity of the problems of economic casuistry which
pious merchants propounded in the confessional. The Spanish dealers on the Antwerp Bourse, a class not morbidly
prone to conscientious scruples, were sufficiently deferential
to ecclesiastical authority to send their confessor to Paris in
order to consult the theologians of the University as to the

compatibility of speculative exchange business with the canon law.[13] When Eck, later famous as the champion who crossed swords with Luther, travelled to Italy, in order to seek from the University of Bologna authoritative confirmation of his daring argument that interest could lawfully be charged in transactions between merchants, no less a group of capitalists than the great house of Fugger thought it worth while to finance an expedition undertaken in quest of so profitable a truth.[14]

Individualistic, competitive, swept forward by an immense expansion of commerce and finance, rather than of industry, and offering opportunities of speculative gain on a scale unknown before, the new economic civilization inevitably gave rise to passionate controversy; and inevitably, since both the friends and the enemies of the Reformation identified it with social change, the leaders in the religious struggle were the protagonists in the debate. In Germany, where social revolution had been fermenting for half a century, it seemed at last to have come. The rise in prices, an enigma which baffled contemporaries till Bodin published his celebrated tract in 1569,[15] produced a storm of indignation against monopolists. Since the rising led by Hans Böheim in 1476, hardly a decade had passed without a peasants' revolt. Usury, long a grievance with craftsman and peasant, had become a battle-cry. From city after city municipal authorities, terrified by popular demands for the repression of the extortioner, consulted universities and divines as to the legitimacy of interest, and universities and divines gave, as is their wont, a loud, but confused, response. Melanchthon expounded godly doctrine on the subject of money-lending and prices.[16] Calvin wrote a famous letter on usury and delivered sermons on the same subject.[17] Bucer sketched a scheme of social reconstruction for a Christian prince.[18] Bullinger produced a classical exposition of social ethics in the *Decades* which he dedicated to Edward VI.[18]

Luther preached and pamphleteered against extortioners,[20] and said that it was time "to put a bit in the mouth of the holy company of the Fuggers." [21]   Zwingli and Œcolampadius devised plans for the reorganization of poor relief.[22] Above all, the Peasants' War, with its touching appeal to the Gospel and its frightful catastrophe, not only terrified Luther into his outburst: "Whoso can, strike, smite, strangle, or stab, secretly or publicly . . . such wonderful times are these that a prince can better merit Heaven with bloodshed than another with prayer"; [23] it also helped to stamp on Lutheranism an almost servile reliance on the secular authorities.   In England there was less violence, but hardly less agitation, and a similar flood of writing and preaching. Latimer, Ponet, Crowley, Lever, Becon, Sandys and Jewel— to mention but the best-known names—all contributed to the debate.[24]   Whatever the social practice of the sixteenth century may have been, it did not suffer for lack of social teaching on the part of men of religion.   If the world could be saved by sermons and pamphlets, it would have been a Paradise.

That the problems of a swiftly changing economic environment should have burst on Europe at a moment when it was torn by religious dissensions more acute than ever before, may perhaps be counted as not least among the tragedies of its history.   But differences of social theory did not coincide with differences of religious opinion, and the mark of nearly all this body of teaching, alike in Germany and in England, is its conservatism.   Where questions of social morality were involved, men whose names are a symbol of religious revolution stood, with hardly an exception, on the ancient ways, appealed to medieval authorities, and reproduced in popular language the doctrines of the Schoolmen.

A view of the social history of the sixteenth century which has found acceptance in certain quarters has represented the Reformation as the triumph of the commercial spirit over the traditional social ethics of Christendom. Something like it is of respectable antiquity. As early as 1540 Cranmer wrote to Oziander protesting against the embarrassment caused to reformers in England by the indulgence to moral laxity, in the matter alike of economic transactions and of marriage, alleged to be given by reformers in Germany.[25] By the seventeenth century the hints had become a theory and an argument. Bossuet taunted Calvin and Bucer with being the first theologians to defend extortion,[26] and it only remained for a pamphleteer to adapt the indictment to popular consumption, by writing bluntly that "it grew to a proverb that usury was the brat of heresy." [27] That the revolt from Rome synchronized, both in Germany and in England, with a period of acute social distress is undeniable, nor is any long argument needed to show that, like other revolutions, it had its seamy side. What is sometimes suggested, however, is not merely a coincidence of religious and economic movements, but a logical connection between changes in economic organization and changes in religious doctrines. It is implied that the bad social practice of the age was the inevitable expression of its religious innovations, and that, if the reformers did not explicitly teach a conscienceless individualism, individualism was, at least, the natural corollary of their teaching. In the eighteenth century, which had as little love for the commercial restrictions of the ages of monkish superstition as for their political theory, that view was advanced as eulogy. In our own day, the wheel seems almost to have come full circle. What was then a matter for congratulation is now often an occasion for criticism. There are writers by whom the Reformation is attacked, as inaugurating a period of unscrupulous commercialism, which had previously been held in check, it is suggested, by the teaching of the Church.

These attempts to relate changes in social theory to the

grand religious struggles of the age have their significance. But the *obiter dicta* of an acrimonious controversy throw more light on the temper of the combatants than on the substance of their contentions, and the issues were too complex to be adequately expressed in the simple antitheses which appealed to partisans. If capitalism means the direction of industry by the owners of capital for their own pecuniary gain, and the social relationships which establish themselves between them and the wage-earning proletariat whom they control, then capitalism had existed on a grand scale both in medieval Italy and in medieval Flanders. If by the capitalist spirit is meant the temper which is prepared to sacrifice all moral scruples to the pursuit of profit, it had been only too familiar to the saints and sages of the Middle Ages. It was the economic imperialism of Catholic Portugal and Spain, not the less imposing, if more solid, achievements of the Protestant powers, which impressed contemporaries down to the Armada. It was predominantly Catholic cities which were the commercial capitals of Europe, and Catholic bankers who were its leading financiers.

Nor is the suggestion that Protestant opinion looked with indulgence on the temper which attacked restraints on economic enterprise better founded. If it is true that the Reformation released forces which were to act as a solvent of the traditional attitude of religious thought to social and economic issues, it did so without design, and against the intention of most reformers. In reality, however sensational the innovations in economic practice which accompanied the expansion of financial capitalism in the sixteenth century, the development of doctrine on the subject of economic ethics was continuous, and, the more closely it is examined, the less foundation does there seem to be for the view that the stream plunged into vacancy over the precipice of the religious revolution. To think of the abdication of religion from its theoretical primacy over economic activity and social institutions as synchronizing with the revolt from Rome, is to antedate a movement which was not finally accomplished for another century and a half, and which owed as much to changes in economic and political organization, as it did to developments in the sphere of religious thought. In the sixteenth century religious teachers of all shades of opinion still searched the Bible, the Fathers and the *Corpus Juris Canonici* for light on practical questions of social morality, and, as far as the first generation of reformers was concerned, there was no intention, among either Lutherans, or Calvinists, or Anglicans, of relaxing the rules of good conscience, which were supposed to control economic transactions and social relations. If anything, indeed, their tendency was to interpret them with a more rigorous severity, as a protest against the moral laxity of the Renaissance, and, in particular, against the avarice which was thought to be peculiarly the sin of Rome. For the passion for regeneration and purification, which was one element in the Reformation, was directed against the corruptions of society as well as of the Church. Princes and nobles and business men conducted themselves after their kind, and fished eagerly in troubled waters. But the aim of religious leaders was to reconstruct, not merely doctrine and ecclesiastical government, but conduct and institutions, on a pattern derived from the forgotten purity of primitive Christianity.

The appeal from the depravity of the present to a golden age of pristine innocence found at once its most vehement, and its most artless, expression in the writings of the German reformers. Like the return to nature in the eighteenth century, it was the cry for spiritual peace of a society disillusioned with the material triumphs of a too complex civilization. The prosperity of Augsburg, Nürnberg, Regensburg, Ulm and Frankfurt, and even of lesser cities like

Rotenburg and Freiburg, had long been the admiration of all observers. Commanding the great trade routes across the Alps and down the Rhine, they had held a central position, which they were to lose when the spice trade moved to Antwerp and Lisbon, and were not to recover till the creation of a railway system in the nineteenth century made Germany again the entrepôt between western Europe and Russia, Austria, Italy and the near East. But the expansion of commerce, which brought affluence to the richer *bourgeoisie,* had been accompanied by the growth of an acute social *malaise,* which left its mark on literature and popular agitation, even before the Discoveries turned Germany from a highway into a back-water. The economic aspect of the development was the rise to a position of overwhelming preëminence of the new interests based on the control of capital and credit. In the earlier Middle Ages capital had been the adjunct and ally of the personal labor of craftsman and artisan. In the Germany of the fifteenth century, as long before in Italy, it had ceased to be a servant and had become a master. Assuming a separate and independent vitality, it claimed the right of a predominant partner to dictate economic organization in accordance with its own exacting requirements.

Under the impact of these new forces, while the institutions of earlier ages survived in form, their spirit and operation were transformed. In the larger cities the gild organization, once a barrier to the encroachments of the capitalist, became one of the instruments which he used to consolidate his power. The rules of fraternities masked a division of the brethren into a plutocracy of merchants, sheltered behind barriers which none but the wealthy craftsman could scale, and a wage-earning proletariat, dependent for their livelihood on capital and credit supplied by their masters, and alternately rising in revolt and sinking in an ever-expanding morass of hopeless pauperism.[28] The peasantry suffered equally from the spread of a commercial civilization into the rural districts and from the survival of ancient agrarian servitudes. As in England, the *nouveaux riches* of the towns invested money in land by purchase and loan, and drove up rents and fines by their competition. But, while in England the customary tenant was shaking off the onerous obligations of villeinage, and appealing, not without success, to the royal courts to protect his title, his brother in south Germany, where serfdom was to last till the middle of the nineteenth century, found *corvées* redoubled, money-payments increased, and common rights curtailed, for the benefit of an impoverished *noblesse,* which saw in the exploitation of the peasant the only means of maintaining its social position in face of the rapidly growing wealth of the *bourgeoisie,* and which seized on the now fashionable Roman law as an instrument to give legal sanction to its harshest exactions.[29]

On a society thus distracted by the pains of growth came the commercial revolution produced by the Discoveries. Their effect was to open a seemingly limitless field to economic enterprise, and to sharpen the edge of every social problem. Unable henceforward to tap through Venice the wealth of the East, the leading commercial houses of south Germany either withdrew from the trade across the Alps, to specialize, like the Fuggers, in banking and finance, or organized themselves into companies, which handled at Lisbon and Antwerp a trade too distant and too expensive to be undertaken by individual merchants using only their own resources. The modern world has seen in America the swift rise of combinations controlling output and prices by the power of massed capital. A somewhat similar movement took place on the narrower stage of European commerce in the generation before the Reformation. Its center was Germany, and it was defended and attacked by arguments almost identical with those which are familiar today.

The exactions of rings and monopolies, which bought in bulk, drove weaker competitors out of the field, "as a great pike swallows up a lot of little fishes," and plundered the consumer, were the commonplaces of the social reformer.[30] The advantages of large-scale organization and the danger of interfering with freedom of enterprise were urged by the companies. The problem was on several occasions brought before the Imperial Diet. But the discovery of the sage who observed that it is not possible to unscramble eggs had already been made, and its decrees, passed in the teeth of strenuous opposition from the interests concerned, do not seem to have been more effective than modern legislation on the same subject.

The passionate anti-capitalist reaction which such conditions produced found expression in numerous schemes of social reconstruction, from the so-called *Reformation of the Emperor Sigismund* in the thirties of the fifteenth century, to the *Twelve Articles* of the peasants in 1525.[31] In the age of the Reformation it was voiced by Hipler, who, in his *Divine Evangelical Reformation,* urged that all merchants' companies, such as those of the Fuggers, Hochstetters and Welsers, should be abolished; by Hutten, who classed merchants with knights, lawyers and the clergy as public robbers; by Geiler von Kaiserberg, who wrote that the monopolists were more detestable than Jews, and should be exterminated like wolves; and, above all, by Luther.[32]

Luther's utterances on social morality are the occasional explosions of a capricious volcano, with only a rare flash of light amid the torrent of smoke and flame, and it is idle to scan them for a coherent and consistent doctrine. Compared with the lucid and subtle rationalism of a thinker like St. Antonino, his sermons and pamphlets on social questions make an impression of naïveté, as of an impetuous but ill-informed genius, dispensing with the cumbrous embarrassments of law and logic, to evolve a system of social ethics from the inspired heat of his own unsophisticated consciousness.

It was partly that they were *pièces de circonstance,* thrown off in the storm of a revolution, partly that it was precisely the refinements of law and logic which Luther detested. Confronted with the complexities of foreign trade and financial organization, or with the subtleties of economic analysis, he is like a savage introduced to a dynamo or a steam-engine. He is too frightened and angry even to feel curiosity. Attempts to explain the mechanism merely enrage him; he can only repeat that there is a devil in it, and that good Christians will not meddle with the mystery of iniquity. But there is a method in his fury. It sprang, not from ignorance, for he was versed in scholastic philosophy, but from a conception which made the learning of the schools appear trivial or mischievous.

"Gold," wrote Columbus, as one enunciating a truism, "constitutes treasure, and he who possesses it has all he needs in this world, as also the means of rescuing souls from Purgatory, and restoring them to the enjoyment of Paradise." [33] It was this doctrine that all things have their price —future salvation as much as present felicity—which scandalized men who could not be suspected of disloyalty to the Church, and which gave their most powerful argument to the reformers. Their outlook on society had this in common with their outlook on religion, that the essence of both was the arraignment of a degenerate civilization before the majestic bar of an uncorrupted past. Of that revolutionary conservatism Luther, who hated the economic individualism of the age not less than its spiritual laxity, is the supreme example. His attitude to the conquest of society by the merchant and financier is the same as his attitude towards the commercialization of religion. When he looks at the Church in Germany, he sees it sucked dry by the tribute which flows to the new Babylon. When he looks at

German social life, he finds it ridden by a conscienceless money-power, which incidentally ministers, like the banking business of the Fuggers, to the avarice and corruption of Rome. The exploitation of the Church by the Papacy, and the exploitation of the peasant and the craftsman by the capitalist, are thus two horns of the beast which sits on the seven hills. Both are essentially pagan, and the sword which will slay both is the same. It is the religion of the Gospel. The Church must cease to be an empire, and become a congregation of believers. Renouncing the prizes and struggles which make the heart sick, society must be converted into a band of brothers, performing in patient cheerfulness the round of simple toil which is the common lot of the descendants of Adam.

The children of the mind are like the children of the body. Once born, they grow by a law of their own being, and, if their parents could foresee their future development, it would sometimes break their hearts. Luther, who has earned eulogy and denunciation as the grand individualist, would have been horrified, could he have anticipated the remoter deductions to be derived from his argument. Wamba said that to forgive as a Christian is not to forgive at all, and a cynic who urged that the Christian freedom expounded by Luther imposed more social restraints than it removed would have more affinity with the thought of Luther himself, than the libertarian who saw in his teaching a plea for treating questions of economic conduct and social organization as spiritually indifferent. Luther's revolt against authority was an attack, not on its rigor, but on its laxity and its corruption. His individualism was not the greed of the plutocrat, eager to snatch from the weakness of public authority an opportunity for personal gain. It was the ingenuous enthusiasm of the anarchist, who hungers for a society in which order and fraternity will reign without "the tedious, stale, forbidding ways of custom, law

and statute," because they well up in all their native purity
from the heart.

Professor Troeltsch has pointed out that Protestants,
not less than Catholics, emphasized the idea of a Church-
civilization, in which all departments of life, the State and
society, education and science, law, commerce and industry, were to be regulated in accordance with the law of
God.[34] That conception dominates all the utterances of
Luther on social issues. So far from accepting the view
which was afterwards to prevail, that the world of business
is a closed compartment with laws of its own, and that the
religious teacher exceeds his commission when he lays down
rules for the moral conduct of secular affairs, he reserves
for that plausible heresy denunciations hardly less bitter
than those directed against Rome. The text of his admonitions is always, "Unless your righteousness exceeds that
of the Scribes and Pharisees," and his appeal is from a formal, legalistic, calculated virtue to the natural kindliness
which does not need to be organized by law, because it is the
spontaneous expression of a habit of love. To restore is to
destroy. The comment on Luther's enthusiasm for the
simple Christian virtues of an age innocent of the artificial
chicaneries of ecclesiastical and secular jurisprudence came
in the thunder of revolution. It was the declaration of the
peasants, that "the message of Christ, the promised Messiah, the word of life, teaching only love, peace, patience
and concord," was incompatible with serfdom, *corvées,* and
enclosures.[35]

The practical conclusion to which such premises led was
a theory of society more medieval than that held by many
thinkers in the Middle Ages, since it dismissed the commercial developments of the last two centuries as a relapse into
paganism. The foundation of it was partly the Bible, partly
a vague conception of a state of nature in which men had
not yet been corrupted by riches, partly the popular pro-

tests against a commercial civilization which were everywhere in the air, and which Luther, a man of the people, absorbed and reproduced with astonishing naïveté, even while he denounced the practical measures proposed to give effect to them.   Like some elements in the Catholic reaction of the twentieth century, the Protestant reaction of the sixteenth sighed for a vanished age of peasant prosperity. The social theory of Luther, who hated commerce and capitalism, has its nearest modern analogy in the Distributive State of Mr. Belloc and Mr. Chesterton.

For the arts by which men amass wealth and power, as for the anxious provision which accumulates for the future, Luther had all the distrust of a peasant and a monk.   Christians should earn their living in the sweat of their brow, take no thought for the morrow, marry young and trust Heaven to provide for its own.   Like Melanchthon, Luther thought that the most admirable life was that of the peasant, for it was least touched by the corroding spirit of commercial calculation, and he quoted Virgil to drive home the lesson to be derived from the example of the patriarchs.[36]   The labor of the craftsman is honorable, for he serves the community in his calling; the honest smith or shoemaker is a priest.   Trade is permissible, provided that it is confined to the exchange of necessaries, and that the seller demands no more than will compensate him for his labor and risk.   The unforgivable sins are idleness and covetousness, for they destroy the unity of the body of which Christians are members.   The grand author and maintainer of both is Rome. For, having ruined Italy, the successor of St. Peter, who lives in a worldly pomp that no king or emperor can equal, has fastened his fangs on Germany; while the mendicant orders, mischievous alike in their practice and by their example, cover the land with a horde of beggars.   Pilgrimages, saints' days and monasteries are an excuse for idleness and must be suppressed.   Vagrants must be either banished or compelled to labor, and each town must organize charity for the support of the honest poor.[37]

Luther accepted the social hierarchy, with its principles of status and subordination, though he knocked away the ecclesiastical rungs in the ladder. The combination of religious radicalism and economic conservatism is not uncommon, and in the traditional conception of society, as an organism of unequal classes with different rights and functions, the father of all later revolutions found an arsenal of arguments against change, which he launched with almost equal fury against revolting peasants and grasping monopolists. His vindication of the spiritual freedom of common men, and his outspoken abuse of the German princes, had naturally been taken at their face value by serfs groaning under an odious tyranny, and, when the inevitable rising came, the rage of Luther, like that of Burke in another age, was sharpened by embarrassment at what seemed to him a hideous parody of truths which were both sacred and his own. As fully convinced as any medieval writer that serfdom was the necessary foundation of society, his alarm at the attempt to abolish it was intensified by a political theory which exalted the absolutism of secular authorities, and a religious doctrine which drew a sharp antithesis between the external order and the life of the spirit. The demand of the peasants that villeinage should end, because "Christ has delivered and redeemed us all, the lowly as well as the great, without exception, by the shedding of His precious blood," [38] horrified him, partly as portending an orgy of anarchy, partly because it was likely to be confused with and to prejudice, as in fact it did, the Reformation movement, partly because (as he thought) it degraded the Gospel by turning a spiritual message into a program of social reconstruction. "This article would make all men equal and so change the spiritual kingdom of Christ into an external worldly one. Impossible! An earthly kingdom cannot exist without inequality of persons. Some must be free, others serfs, some rulers, others subjects. As St. Paul says, 'Before Christ both master and slave are one.' " [39] After nearly four centuries, Luther's apprehensions of a too hasty establishment of the Kingdom of Heaven appear somewhat exaggerated.

A society may perish by corruption as well as by violence. Where the peasants battered, the capitalist mined; and Luther, whose ideal was the patriarchal ethics of a world which, if it ever existed, was visibly breaking up, had as little mercy for the slow poison of commerce and finance as for the bludgeon of revolt. No contrast could be more striking than that between his social theory and the outlook of Calvin. Calvin, with all his rigor, accepted the main institutions of a commercial civilization, and supplied a creed to the classes which were to dominate the future. The eyes of Luther were on the past. He saw no room in a Christian society for those middle classes whom an English statesman once described as the natural representatives of the human race. International trade, banking and credit, capitalist industry, the whole complex of economic forces, which, next to his own revolution, were to be the mightiest solvent of the medieval world, seem to him to belong in their very essence to the kingdom of darkness which the Christian will shun. He attacks the authority of the canon law, only to reaffirm more dogmatically the detailed rules which it had been used to enforce. When he discusses economic questions at length, as in his *Long Sermon on Usury* in 1520, or his tract *On Trade and Usury* in 1524, his doctrines are drawn from the straitest interpretation of ecclesiastical jurisprudence, unsoftened by the qualifications with which canonists themselves had attempted to adapt its rigors to the exigencies of practical life.

In the matter of prices he merely rehearses traditional

doctrines. "A man should not say, 'I will sell my wares as dear as I can or please,' but 'I will sell my wares as is right and proper.' For thy selling should not be a work that is within thy own power or will, without all law and limit, as though thou wert a God, bounden to no one. But because thy selling is a work that thou performest to thy neighbor, it should be restrained within such law and conscience that thou mayest practice it without harm or injury to him." [40] If a price is fixed by public authority, the seller must keep to it. If it is not, he must follow the price of common estimation. If he has to determine it himself, he must consider the income needed to maintain him in his station in life, his labor, and his risk, and must settle it accordingly. He must not take advantage of scarcity to raise it. He must not corner the market. He must not deal in futures. He must not sell dearer for deferred payments.

On the subject of usury, Luther goes even further than the orthodox teaching. He denounces the concessions to practical necessities made by the canonists. "The greatest misfortune of the German nation is easily the traffic in interest. . . . The devil invented it, and the Pope, by giving his sanction to it, has done untold evil throughout the world." [41] Not content with insisting that lending ought to be free, he denounces the payment of interest as compensation for loss and the practice of investing in rent-charges, both of which the canon law in his day allowed, and would refuse usurers the sacrament, absolution, and Christian burial. With such a code of ethics, Luther naturally finds the characteristic developments of his generation—the luxury trade with the East, international finance, speculation on the exchanges, combinations and monopolies—shocking beyond measure. "Foreign merchandise which brings from Calicut and India and the like places wares such as precious silver and jewels and spices . . . and drain the land and people of their money, should not be permitted. . . . Of combinations I ought really to say much, but the matter is endless and bottomless, full of mere greed and wrong. . . . Who is so stupid as not to see that combinations are mere outright monopolies, which even heathen civil laws—I will say nothing of divine right and Christian law—condemn as a plainly harmful thing in all the world?" [42]

So resolute an enemy'of license might have been expected to be the champion of law.  It might have been supposed that Luther, with his hatred of the economic appetites, would have hailed as an ally the restraints by which, at least in theory, those appetites had been controlled.  In reality, of course, his attitude towards the mechanism of ecclesiastical jurisprudence and discipline was the opposite.  It was one, not merely of indifference, but of repugnance.  The prophet who scourged with whips the cupidity of the individual chastised with scorpions the restrictions imposed upon it by society; the apostle of an ideal ethic of Christian love turned a shattering dialectic on the corporate organization of the Christian Church.  In most ages, so tragic a parody of human hopes are human institutions, there have been some who have loved mankind, while hating almost everything that men have done or made.  Of that temper Luther, who lived at a time when the contrast between a sublime theory and a hideous reality had long been intolerable, is the supreme example.  He preaches a selfless charity, but he recoils with horror from every institution by which an attempt had been made to give it a concrete expression.  He reiterates the content of medieval economic teaching with a literalness rarely to be found in the thinkers of the later Middle Ages, but for the rules and ordinances in which it had received a positive, if sadly imperfect, expression, he has little but abhorrence.  God speaks to the soul, not through the mediation of the priesthood or of social institutions built up by man, but *solus cum solo,* as a voice in the heart and in the heart alone.  Thus the bridges between the worlds of spirit and of sense are broken, and the soul is isolated from the society of men, that it may enter into communion with its Maker. The grace that is freely bestowed upon it may overflow in its social relations; but those relations can supply no particle of spiritual nourishment to make easier the reception of grace. Like the primeval confusion into which the fallen Angel plunged on his fatal mission, they are a chaos of brute matter, a wilderness of dry bones, a desert unsanctified and incapable of contributing to sanctification. "It is certain that absolutely none among outward things, under whatever name they may be reckoned, has any influence in producing Christian righteousness or liberty. . . . One thing, and one alone, is necessary for life, justification and Christian liberty; and that is the most holy word of God, the Gospel of Christ." [43]

The difference between loving men as a result of first loving God, and learning to love God through a growing love for men, may not, at first sight, appear profound. To Luther it seemed an abyss, and Luther was right. It was, in a sense, nothing less than the Reformation itself. For carried, as it was not carried by Luther, to its logical result, the argument made, not only good works, but sacraments and the Church itself unnecessary. The question of the religious significance of that change of emphasis, and of the validity of the intellectual processes by which Luther reached his conclusions, is one for theologians. Its effects on social theory were staggering. Since salvation is bestowed by the operation of grace in the heart and by that alone, the whole fabric of organized religion, which had mediated between the individual soul and its Maker—divinely commissioned hierarchy, systematized activities, corporate institutions—drops away, as the blasphemous trivialities of a religion of works. The medieval conception of the social order, which had regarded it as a highly articulated organism of members contributing in their different degrees to a spiritual purpose, was shattered, and differences which had been distinctions within a larger unity were now set in irreconcilable antagonism to each other. Grace no longer completed nature: it was the antithesis of it. Man's actions as a member of society were no longer the extension of his life as a child of God: they were its negation. Secular interests ceased to possess, even remotely, a religious significance: they might compete with religion, but they could not enrich it. Detailed rules of conduct—a Christian casuistry—are needless or objectionable: the Christian has a sufficient guide in the Bible and in his own conscience. In one sense, the distinction between the secular and the religious life vanished. Monasticism was, so to speak, secularized; all men stood henceforward on the same footing towards God; and that advance, which contained the germ of all subsequent revolutions, was so enormous that all else seems insignificant. In another sense, the distinction became more profound than ever before. For, though all might be sanctified, it was their inner life alone which could partake of sanctification. The world was divided into good and evil, light and darkness, spirit and matter. The division between them was absolute; no human effort could span the chasm.

The remoter corollaries of the change remained to be stated by subsequent generations. Luther himself was not consistent. He believed that it was possible to maintain the content of medieval social teaching, while rejecting its sanctions, and he insisted that good works would be the fruit of salvation as vehemently as he denied that they could contribute to its attainment. In his writings on social questions emphasis on the traditional Christian morality is combined with a repudiation of its visible and institutional framework, and in the tragic struggle which results between spirit and letter, form and matter, grace and works, his intention, at least, is not to jettison the rules of good conscience in economic matters, but to purify them by an immense effort of simplification. His denunciation of medieval charity, fraternities, mendicant orders, festivals and pilgrimages, while it drew its point from practical abuses, sprang inevitably from his repudiation of the idea that merit could be acquired by the operation of some special machinery beyond the conscientious discharge of the ordinary duties of daily life. His demand for the abolition of the canon law was the natural corollary of his belief that the Bible was an all-sufficient guide to action. While not rejecting ecclesiastical discipline altogether, he is impatient of it. The Christian, he argues, needs no elaborate mechanism to teach him his duty or to correct him if he neglects it. He has the Scriptures and his own conscience; let him listen to *them*. "There can be no better instructions in . . . all transactions in temporal goods than that every man who is to deal with his neighbor present to himself these commandments: 'What ye would that others should do unto you, do ye also unto them,' and 'Love thy neighbor as thyself.' If these were followed out, then everything would instruct and arrange itself; then no law books nor courts nor judicial actions would be required; all things would quietly and simply be set to rights, for every one's heart and conscience would guide him." [44]

"Everything would arrange itself." Few would deny it. But how if it does not? Is emotion really an adequate substitute for reason, and rhetoric for law? Is it possible to solve the problem which social duties present to the individual by informing him that no problem exists? If it is true that the inner life is the sphere of religion, does it necessarily follow that the external order is simply irrelevant to it? To wave aside the world of institutions and law as alien to that of the spirit—is not this to abandon, instead of facing, the task of making Christian morality prevail, for which medieval writers, with their conception of a hierarchy of values related to a common end, had attempted, however inadequately, to discover a formula? A Catholic rationalist had answered by anticipation Luther's contemptuous dismissal of law and learning, when he urged that it was useless for the Church to prohibit extortion, unless it was prepared to undertake the intellectual labor of defining the transactions to which the prohibition applied.[45]  It was a pity that Pecock's douche of common sense was not of a kind which could be appreciated by Luther.  He denounced covetousness in general terms, with a surprising exuberance of invective.  But, confronted with a request for advice on the specific question whether the authorities of Dantzig shall put down usury, he retreats into the clouds.  "The preacher shall preach only the Gospel rule, and leave it to each man to follow his own conscience.  Let him who can receive it, receive it; he cannot be compelled thereto further than the Gospel leads willing hearts whom the spirit of God urges forward." [46]

Luther's impotence was not accidental.  It sprang directly from his fundamental conception that to externalize religion in rules and ordinances is to degrade it.  He attacked the casuistry of the canonists, and the points in their teaching with regard to which his criticism was justified were only too numerous.  But the remedy for bad law is good law, not lawlessness; and casuistry is merely the application of general principles to particular cases, which is involved in any living system of jurisprudence, whether ecclesiastical or secular.  If the principles are not to be applied, on the ground that they are too sublime to be soiled by contact with the gross world of business and politics, what remains of them?  Denunciations such as Luther launched against the Fuggers and the peasants; aspirations for an idyll of Christian charity and simplicity, such as he advanced in his tract *On Trade and Usury*.  Pious rhetoric

may be edifying, but it is hardly the panoply recommended
by St. Paul.

"As the soul needs the word alone for life and justification, so it is justified by faith alone, and not by any works.
. . . Therefore the first care of every Christian ought to
be to lay aside all reliance on works, and to strengthen his
faith alone more and more." [47] The logic of Luther's religious premises was more potent for posterity than his attachment to the social ethics of the past, and evolved its own
inexorable conclusions in spite of them.   It enormously
deepened spiritual experience, and sowed the seeds from
which new freedoms, abhorrent to Luther, were to spring.
But it riveted on the social thought of Protestantism a dualism which, as its implications were developed, emptied religion of its social content, and society of its soul.   Between
light and darkness a great gulf was fixed.   Unable to climb
upwards plane by plane, man must choose between salvation
and damnation.   If he despairs of attaining the austere
heights where alone true faith is found, no human institution can avail to help him.   Such, Luther thinks, will be the
fate of only too many.

He himself was conscious that he had left the world of
secular activities perilously divorced from spiritual restraints.   He met the difficulty, partly with an admission
that it was insuperable, as one who should exult in the majestic unreasonableness of a mysterious Providence, whose
decrees might not be broken, but could not, save by a few,
be obeyed; partly with an appeal to the State to occupy the
province of social ethics, for which his philosophy could
find no room in the Church.   "Here it will be asked, 'Who
then can be saved, and where shall we find Christians?   For
in this fashion no merchandising would remain on earth.'
. . . You see it is as I said, that Christians are rare people
on earth.   Therefore stern hard civil rule is necessary in
the world, lest the world become wild, peace vanish, and

commerce and common interests be destroyed. . . . No one need think that the world can be ruled without blood. The civil sword shall and must be red and bloody." [48]

Thus the axe takes the place of the stake, and authority, expelled from the altar, finds a new and securer home upon the throne. The maintenance of Christian morality is to be transferred from the discredited ecclesiastical authorities to the hands of the State. Skeptical as to the existence of unicorns and salamanders, the age of Machiavelli and Henry VIII found food for its credulity in the worship of that rare monster, the God-fearing Prince.

### III. CALVIN

The most characteristic and influential form of Protestantism in the two centuries following the Reformation is that which descends, by one path or another, from the teaching of Calvin. Unlike the Lutheranism from which it sprang, Calvinism, assuming different shapes in different countries, became an international movement, which brought, not peace, but a sword, and the path of which was strewn with revolutions. Where Lutheranism had been socially conservative, deferential to established political authorities, the exponent of a personal, almost a quietistic, piety, Calvinism was an active and radical force. It was a creed which sought, not merely to purify the individual, but to reconstruct Church and State, and to renew society by penetrating every department of life, public as well as private, with the influence of religion.

Upon the immense political reactions of Calvinism, this is not the place to enlarge. As a way of life and a theory of society, it possessed from the beginning one characteristic which was both novel and important. It assumed an economic organization which was relatively advanced, and expounded its social ethics on the basis of it. In this respect the teaching of the Puritan moralists who derive most directly from Calvin is in marked contrast with that both of medieval theologians and of Luther. The difference is not merely one of the conclusions reached, but of the plane on which the discussion is conducted. The background, not only of most medieval social theory, but also of Luther and his English contemporaries, is the traditional stratification of rural society. It is a natural, rather than a money, economy, consisting of the petty dealings of peasants and craftsmen in the small market town, where industry is carried on for the subsistence of the household and the consumption of wealth follows hard upon the production of it, and where commerce and finance are occasional incidents, rather than the forces which keep the whole system in motion. When they criticize economic abuses, it is precisely against departures from that natural state of things —against the enterprise, the greed of gain, the restless competition, which disturb the stability of the existing order with clamorous economic appetites—that their criticism is directed.

These ideas were the traditional retort to the evils of unscrupulous commercialism, and they left some trace on the writings of the Swiss reformers. Zwingli, for example, who, in his outlook on society, stood midway between Luther and Calvin, insists on the oft-repeated thesis that private property originates in sin; warns the rich that they can hardly enter the Kingdom of Heaven; denounces the Councils of Constance and Basel—"assembled, forsooth, at the bidding of the Holy Ghost"—for showing indulgence to the mortgaging of land on the security of crops; and, while emphasizing that interest must be paid when the State sanctions it, condemns it in itself as contrary to the law of God.[49] Of the attempts made at Zürich and Geneva to repress extortion something is said below. But these full-blooded denunciations of capitalism were not intended by their authors to supply a rule of practical life, since it was the duty of the individual to comply with the secular legislation by which interest was permitted, and already, when they were uttered, they had ceased to represent the conclusion of the left wing of the Reformed Churches.

For Calvin, and still more his later interpreters, began their voyage lower down the stream. Unlike Luther, who saw economic life with the eyes of a peasant and a mystic, they approached it as men of affairs, disposed neither to idealize the patriarchal virtues of the peasant community, nor to regard with suspicion the mere fact of capitalist enterprise in commerce and finance. Like early Christianity and modern socialism, Calvinism was largely an urban movement; like them, in its earlier days, it was carried from country to country partly by emigrant traders and workmen; and its stronghold was precisely in those social groups to which the traditional scheme of social ethics, with its treatment of economic interests as a quite minor aspect of human affairs, must have seemed irrelevant or artificial. As was to be expected in the exponents of a faith which had its headquarters at Geneva, and later its most influential adherents in great business centers, like Antwerp with its industrial hinterland, London, and Amsterdam, its leaders addressed their teaching, not of course exclusively, but none the less primarily, to the classes engaged in trade and industry, who formed the most modern and progressive elements in the life of the age.

In doing so they naturally started from a frank recognition of the necessity of capital, credit and banking, large-scale commerce and finance, and the other practical facts of business life. They thus broke with the tradition which, regarding a preoccupation with economic interests "beyond what is necessary for subsistence" as reprehensible, had stigmatized the middleman as a parasite and the usurer as a thief. They set the profits of trade and finance, which to

the medieval writer, as to Luther, only with difficulty escaped censure as *turpe lucrum,* on the same level of respectability as the earnings of the laborer and the rents of
the landlord. "What reason is there," wrote Calvin to a
correspondent, "why the income from business should not
be larger than that from land-owning? Whence do the
merchant's profits come, except from his own diligence and
industry?" [50] It was quite in accordance with the spirit of
those words that Bucer, even while denouncing the frauds
and avarice of merchants, should urge the English Government to undertake the development of the woollen industry on mercantilist lines.[51]

Since it is the environment of the industrial and commercial classes which is foremost in the thoughts of Calvin
and his followers, they have to make terms with its practical necessities. It is not that they abandon the claim of
religion to moralize economic life, but that the life which
they are concerned to moralize is one in which the main
features of a commercial civilization are taken for granted,
and that it is for application to such conditions that their
teaching is designed. Early Calvinism, as we shall see, has
its own rule, and a rigorous rule, for the conduct of economic affairs. But it no longer suspects the whole world of
economic motives as alien to the life of the spirit, or dis-
trusts the capitalist as one who has necessarily grown rich
on the misfortunes of his neighbor, or regards poverty as
in itself meritorious, and it is perhaps the first systematic
body of religious teaching which can be said to recognize
and applaud the economic virtues. Its enemy is not the accumulation of riches, but their misuse for purposes of self-
indulgence or ostentation. Its ideal is a society which seeks
wealth with the sober gravity of men who are conscious at
once of disciplining their own characters by patient labor,
and of devoting themselves to a service acceptable to God.

It is in the light of that change of social perspective that the doctrine of usury associated with the name of Calvin is to be interpreted. Its significance consisted, not in the phase which it marked in the technique of economic analysis, but in its admission to a new position of respectability of a powerful and growing body of social interests, which, however irrepressible in practice, had hitherto been regarded by religious theory as, at best, of dubious propriety, and, at worst, as frankly immoral. Strictly construed, the famous pronouncement strikes the modern reader rather by its rigor than by its indulgence. "Calvin," wrote an English divine a generation after his death, "deals with usurie as the apothecarie doth with poyson." [52] The apologetic was just, for neither his letter to Œcolampadius, nor his sermon on the same subject, reveal any excessive tolerance for the trade of the financier. That interest is lawful, provided that it does not exceed an official maximum, that, even when a maximum is fixed, loans must be made *gratis* to the poor, that the borrower must reap as much advantage as the lender, that excessive security must not be exacted, that what is venial as an occasional expedient is reprehensible when carried on as a regular occupation, that no man may snatch economic gain for himself to the injury of his neighbor—a condonation of usury protected by such embarrassing entanglements can have offered but tepid consolation to the devout moneylender.

Contemporaries interpreted Calvin to mean that the debtor might properly be asked to concede some small part of his profits to the creditor with whose capital they had been earned, but that the exaction of interest was wrong if it meant that "the creditor becomes rich by the sweat of the debtor, and the debtor does not reap the reward of his labor." There have been ages in which such doctrines would have been regarded as an attack on financial enterprise rather than as a defense of it. Nor were Calvin's specific contributions to the theory of usury strikingly original. As a hard-headed lawyer, he was free both from the incoherence and from the idealism of Luther, and his doctrine was probably regarded by himself merely as one additional step in the long series of developments through which ecclesiastical jurisprudence on the subject had already gone. In emphasizing the difference between the interest wrung from the necessities of the poor and the interest which a prosperous merchant could earn with borrowed capital, he had been anticipated by Major; in his sanction of a moderate rate on loans to the rich, his position was the same as that already assumed, though with some hesitation, by Melanchthon. The picture of Calvin, the organizer and disciplinarian, as the parent of laxity in social ethics, is a legend. Like the author of another revolution in economic theory, he might have turned on his popularizers with the protest: "I am not a Calvinist."

Legends are apt, however, to be as right in substance as they are wrong in detail, and both its critics and its defenders were correct in regarding Calvin's treatment of capital as a watershed. What he did was to change the plane on which the discussion was conducted, by treating the ethics of money-lending, not as a matter to be decided by an appeal to a special body of doctrine on the subject of usury, but as a particular case of the general problem of the social relations of a Christian community, which must be solved in the light of existing circumstances. The significant feature in his discussion of the subject is that he assumes credit to be a normal and inevitable incident in the life of society. He therefore dismisses the oft-quoted passages from the Old Testament and the Fathers as irrelevant, because designed for conditions which no longer exist, argues that the payment of interest for capital is as reasonable as the payment of rent for land, and throws on the conscience of the individual the obligation of seeing that it does not exceed the amount dictated by natural justice and the golden rule. He makes, in short, a fresh start, argues that what is permanent is, not the rule *"non fœnerabis,"* but *"l'équité et la droiture,"* and appeals from Christian tradition to commercial common sense, which he is sanguine enough to hope will be Christian. On such a view all extortion is to be avoided by Christians. But capital and credit are indispensable; the financier is not a pariah, but a useful member of society; and lending at interest, provided that the rate is reasonable and that loans are made freely to the poor, is not *per se* more extortionate than any other of the economic transactions without which human affairs cannot be carried on. That acceptance of the realities of commercial practice as a starting-point was of momentous importance. It meant that Calvinism and its off-shoots took their stand on the side of the activities which were to be most characteristic of the future, and insisted that it was not by renouncing them, but by untiring concentration on the task of using for the glory of God the opportunities which they offered, that the Christian life could and must be lived.

It was on this practical basis of urban industry and commercial enterprise that the structure of Calvinistic social ethics was erected. Upon their theological background it would be audacious to enter. But even an amateur may be pardoned, if he feels that there have been few systems in which the practical conclusions flow by so inevitable a logic from the theological premises. "God not only foresaw," Calvin wrote, "the fall of the first man, . . . but also arranged all by the determination of his own will." [53] Certain individuals he chose as his elect, predestined to salvation from eternity by "his gratuitous mercy, totally irrespective of human merit"; the remainder have been consigned to eternal damnation, "by a just and irreprehensible, but incomprehensible, judgment." [54] Deliverance, in short, is the work, not of man himself, who can contribute nothing to it, but of an objective Power. Human effort, social institutions, the world of culture, are at best irrelevant to salvation, and at worst mischievous. They distract man from the true aim of his existence and encourage reliance upon broken reeds.

That aim is not personal salvation, but the glorification of God, to be sought, not by prayer only, but by action—the sanctification of the world by strife and labor. For Calvinism, with all its repudiation of personal merit, is intensely practical. Good works are not a way of attaining salvation, but they are indispensable as a proof that salvation has been attained. The central paradox of religious ethics—that only those are nerved with the courage needed to turn the world upside down, who are convinced that already, in a higher sense, it is disposed for the best by a Power of which they are the humble instruments—finds in it a special exemplification. For the Calvinist the world is ordained to show forth the majesty of God, and the duty of the Christian is to live for that end. His task is at once to discipline his individual life, and to create a sanctified society. The Church, the State, the community in which he lives, must not merely be a means of personal salvation, or minister to his temporal needs. It must be a "Kingdom of Christ," in which individual duties are performed by men conscious that they are "ever in their great Taskmaster's eye," and the whole fabric is preserved from corruption by a stringent and all-embracing discipline.

The impetus to reform or revolution springs in every age from the realization of the contrast between the external order of society and the moral standards recognized as valid by the conscience or reason of the individual. And naturally it is in periods of swift material progress, such as the sixteenth and eighteenth centuries, that such a contrast is most acutely felt. The men who made the Reformation had seen the Middle Ages close in the golden autumn which, amid all the corruption and tyranny of the time, still glows in the pictures of Nürnberg and Frankfurt drawn by Æneas Silvius and in the woodcuts of Dürer. And already a new dawn of economic prosperity was unfolding. Its promise was splendid, but it had been accompanied by a cynical materialism which seemed a denial of all that had been meant by the Christian virtues, and which was the more horrifying because it was in the capital of the Christian Church that it reached its height. Shocked by the gulf between theory and practice, men turned this way and that to find some solution of the tension which racked them. The German reformers followed one road and preached a return to primitive simplicity. But who could obliterate the achievements of two centuries, or blot out the new worlds which science had revealed? The Humanists took another, which should lead to the gradual regeneration of mankind by the victory of reason over superstition and brutality and avarice. But who could wait for so distant a consummation? Might there not be a third? Was it not possible that, purified and disciplined, the very qualities which economic success demanded—thrift, diligence, sobriety, frugality—were themselves, after all, the foundation, at least, of the Christian virtues? Was it not conceivable that the gulf which yawned between a luxurious world and the life of the spirit could be bridged, not by eschewing material interests as the kingdom of darkness, but by dedicating them to the service of God?

It was that revolution in the traditional scale of ethical values which the Swiss reformers desired to achieve; it was that new type of Christian character that they labored to create. Not as part of any scheme of social reform, but as elements in a plan of moral regeneration, they seized on the aptitudes cultivated by the life of business and affairs, stamped on them a new sanctification, and used them as the warp of a society in which a more than Roman discipline should perpetuate a character the exact antithesis of that fostered by obedience to Rome. The Roman Church, it was held, through the example of its rulers, had encouraged luxury and ostentation: the members of the Reformed Church must be economical and modest. It had sanctioned the spurious charity of indiscriminate almsgiving: the true Christian must repress mendicancy and insist on the virtues of industry and thrift. It had allowed the faithful to believe that they could atone for a life of worldliness by the savorless formality of individual good works reduced to a commercial system, as though man could keep a profit and loss account with his Creator: the true Christian must organize his life as a whole for the service of his Master. It had rebuked the pursuit of gain as lower than the life of religion, even while it took bribes from those who pursued gain with success: the Christian must conduct his business with a high seriousness, as in itself a kind of religion.

Such teaching, whatever its theological merits or defects, was admirably designed to liberate economic energies, and to weld into a disciplined social force the rising *bourgeoisie,* conscious of the contrast between its own standards and those of a laxer world, proud of its vocation as the standard-bearer of the economic virtues, and determined to vindicate an open road for its own way of life by the use of every weapon, including political revolution and war, because the issue which was at stake was not merely convenience or self-interest, but the will of God. Calvinism stood, in short, not only for a new doctrine of theology and ecclesiastical government, but for a new scale of moral values and a new ideal of social conduct. Its practical message, it might perhaps be said, was *la carrière ouverte*—not *aux talents,* but *au caractère.*

Once the world had been settled to their liking, the middle classes persuaded themselves that they were the convinced enemies of violence and the devotees of the principle of order. While their victories were still to win, they were everywhere the spear-head of revolution. It is not wholly fanciful to say that, on a narrower stage but with not less formidable weapons, Calvin did for the *bourgeoisie* of the sixteenth century what Marx did for the proletariat of the nineteenth, or that the doctrine of predestination satisfied the same hunger for an assurance that the forces of the universe are on the side of the elect as was to be assuaged in a different age by the theory of historical materialism. He set their virtues at their best in sharp antithesis with the vices of the established order at its worst, taught them to feel that they were a chosen people, made them conscious of their great destiny in the Providential plan and resolute to realize it. The new law was graven on tablets of flesh; it not merely rehearsed a lesson, but fashioned a soul. Compared with the quarrelsome, self-indulgent nobility of most European countries, or with the extravagant and half-bankrupt monarchies, the middle classes, in whom Calvinism took root most deeply, were a race of iron. It was not surprising that they made several revolutions, and imprinted their conceptions of political and social expediency on the public life of half a dozen different States in the Old World and in the New.

The two main elements in this teaching were the insistence on personal responsibility, discipline and asceticism, and the call to fashion for the Christian character an objective embodiment in social institutions. Though logically connected, they were often in practical discord. The influence of Calvinism was not simple, but complex, and extended far beyond the circle of Churches which could properly be called Calvinist. Calvinist theology was accepted where Calvinist discipline was repudiated. The bitter struggle between Presbyterians and Independents in England did not prevent men, to whom the whole idea of religious uniformity was fundamentally abhorrent, from drawing inspiration from the conception of a visible Christian society, in which, as one of them said, the Scripture was "really and materially to be fulfilled." [55] Both an intense individualism and a rigorous Christian Socialism could be deduced from Calvin's doctrine. Which of them predominated depended on differences of political environment and of social class. It depended, above all, on the question whether Calvinists were, as at Geneva and in Scotland, a majority, who could stamp their ideals on the social order, or, as in England, a minority, living on the defensive beneath the suspicious eyes of a hostile Government.

In the version of Calvinism which found favor with the English upper classes in the seventeenth century, individualism in social affairs was, on the whole, the prevalent philosophy. It was only the fanatic and the agitator who drew inspiration from the vision of a New Jerusalem descending on England's green and pleasant land, and the troopers of Fairfax soon taught them reason. But, if the theology of Puritanism was that of Calvin, its conception of society, diluted by the practical necessities of a commercial age, and softened to suit the conventions of a territorial aristocracy, was poles apart from that of the master who founded a discipline, compared with which that of Laud, as Laud himself dryly observed,[56] was a thing of shreds and patches. As both the teaching of Calvin himself, and the practice of some Calvinist communities, suggest, the social ethics of the heroic age of Calvinism savored more of a collectivist dictatorship than of individualism. The expression of a revolt against the medieval ecclesiastical system, it stood itself, where circumstances favored it, for a discipline far more stringent and comprehensive than that of the Middle Ages. If, as some historians have argued, the philosophy of *laissez faire* emerged as a result of the spread of Calvinism among the middle classes, it did so, like tolerance, by a route which was indirect. It was accepted, less because it was

esteemed for its own sake, than as a compromise forced
upon Calvinism at a comparatively late stage in its history,
as a result of its modification by the pressure of commercial
interests, or of a balance of power between conflicting authorities.

The spirit of the system is suggested by its treatment of
the burning question of Pauperism. The reform of traditional methods of poor relief was in the air—Vives had
written his celebrated book in 1526 [57]—and, prompted both
by Humanists and by men of religion, the secular authorities all over Europe were beginning to bestir themselves to
cope with what was, at best, a menace to social order, and,
at worst, a moral scandal. The question was naturally one
which appealed strongly to the ethical spirit of the Reformation. The characteristic of the Swiss reformers, who were
much concerned with it, was that they saw the situation not,
like the statesman, as a problem of police, nor, like the more
intelligent Humanists, as a problem of social organization,
but as a question of character. Calvin quoted with approval
the words of St. Paul, "If a man will not work, neither shall
he eat," condemned indiscriminate alms-giving as vehemently as any Utilitarian, and urged that the ecclesiastical
authorities should regularly visit every family to ascertain
whether its members were idle, or drunken, or otherwise un-
desirable.[58] Œcolampadius wrote two tracts on the relief of
the poor.[59] Bullinger lamented the army of beggars produced by monastic charity, and secured part of the emolu-
ments of a dissolved abbey for the maintenance of a school
and the assistance of the destitute.[60] In the plan for the reorganization of poor relief at Zürich, which was drafted by
Zwingli in 1525, all mendicancy was strictly forbidden;
travellers were to be relieved on condition that they left the
town next day; provision was to be made for the sick and
aged in special institutions; no inhabitant was to be entitled
to relief who wore ornaments or luxurious clothes, who

failed to attend church, or who played cards or was otherwise disreputable. The basis of his whole scheme was the
duty of industry and the danger of relaxing the incentive
to work. "With labor," he wrote, "will no man now support himself. . . . And yet labor is a thing so good and
godlike . . . that makes the body hale and strong and
cures the sicknesses produced by idleness. . . . In the
things of this life, the laborer is most like to God." [61]

In the assault on pauperism, moral and economic motives
were not distinguished. The idleness of the mendicant was
both a sin against God and a social evil; the enterprise of
the thriving tradesman was at once a Christian virtue and
a benefit to the community. The same combination of religious zeal and practical shrewdness prompted the attacks
on gambling, swearing, excess in apparel and self-indulgence in eating and drinking. The essence of the system
was not preaching or propaganda, though it was prolific
of both, but the attempt to crystallize a moral ideal in the
daily life of a visible society, which should be at once a
Church and a State. Having overthrown monasticism, its
aim was to turn the secular world into a gigantic monastery,
and at Geneva, for a short time, it almost succeeded. "In
other places," wrote Knox of that devoted city, "I confess
Christ to be truly preached, but manners and religion so sincerely reformed I have not yet seen in any place besides." [62]
Manners and morals were regulated, because it is through
the *minutiæ* of conduct that the enemy of mankind finds his
way to the soul; the traitors to the Kingdom might be revealed by pointed shoes or golden ear-rings, as in 1793 those
guilty of another kind of *incivisme* were betrayed by their
knee-breeches. Regulation meant legislation, and, still
more, administration. The word in which both were summarized was Discipline.

Discipline Calvin himself described as the nerves of religion,[63] and the common observation that he assigned to it

the same primacy as Luther had given to faith is just. As organized in the Calvinist Churches, it was designed primarily to safeguard the sacrament and to enforce a censorship of morals, and thus differed in scope and purpose from the canon law of the Church of Rome, as the rules of a private society may differ from the code of a State. Its establishment at Geneva, in the form which it assumed in the last half of the sixteenth century, was the result of nearly twenty years of struggle between the Council of the city and the Consistory, composed of ministers and laymen. It was only in 1555 that the latter finally vindicated its right to excommunicate, and only in the edition of the *Institutes* which appeared in 1559 that a scheme of church organization and discipline was set out. But, while the answer to the question of the constitution of the authority by whom discipline was to be exercised depended on political conditions, and thus differed in different places and periods, the necessity of enforcing a rule of life, which was the practical aspect of discipline, was from the start of the very essence of Calvinism. Its importance was the theme of a characteristic letter addressed by Calvin to Somerset in October 1548, the moment of social convulsion for which Bucer wrote his book, *De Regno Christi*. The Protector is reminded that it is not from lack of preaching, but from failure to enforce compliance with it, that the troubles of England have sprung. Though crimes of violence are punished, the licentious are spared, and the licentious have no part in the Kingdom of God. He is urged to make sure that "les hommes soient tenus en bonne et honneste discipline," and to be careful "que ceulx qui oyent la doctrine de l'Evangile s'approuvent estre Chrestiens par sainctité de vie." [64]

"Prove themselves Christians by holiness of life"—the words might be taken as the motto of the Swiss reformers, and their projects of social reconstruction are a commentary on the sense in which "holiness of life" was understood.

It was in that spirit that Zwingli took the initiative in form-
ing at Zürich a board of moral discipline, to be composed of
the clergy, the magistrates and two elders; emphasized the
importance of excommunicating offenders against Christian
morals; and drew up a list of sins to be punished by excommunication, which included, in addition to murder and
theft, unchastity, perjury and avarice, "especially as it discovers itself in usury and fraud." [65] It was in that spirit
that Calvin composed in the *Institutes* a Protestant *Summa*
and manual of moral casuistry, in which the lightest action
should be brought under the iron control of a universal rule.
It was in that spirit that he drafted the heads of a comprehensive scheme of municipal government, covering the whole
range of civic administration, from the regulations to be
made for markets, crafts, buildings and fairs to the control
of prices, interest and rents.[66] It was in that spirit that he
made Geneva a city of glass, in which every household lived
its life under the supervision of a spiritual police, and that
for a generation Consistory and Council worked hand in
hand, the former excommunicating drunkards, dancers and
contemners of religion, the latter punishing the dissolute
with fines and imprisonment and the heretic with death.
"Having considered," ran the preamble to the ordinances
of 1576, which mark the maturity of the Genevese Church,
"that it is a thing worthy of commendation above all others,
that the doctrine of the Holy Gospel of our Lord Jesus
Christ shall be preserved in its purity, and the Christian
Church duly maintained by good government and policy,
and also that youth in the future be well and faithfully instructed, and the Hospital well ordered for the support of
the poor: Which things can only be if there be established
a certain rule and order of living, by which each man may
be able to understand the duties of his position. . . ." [67]
The object of it all was so simple. "Each man to understand the duties of his position"—what could be more de-

sirable, at Geneva or elsewhere? It is sad to reflect that the attainment of so laudable an end involved the systematic use of torture, the beheading of a child for striking its parents, and the burning of a hundred and fifty heretics in sixty years.[68] *Tantum religio potuit suadere malorum.*

Torturing and burning were practised elsewhere by Governments which affected no excessive zeal for righteousness. The characteristic which was distinctive of Geneva—"the most perfect school of Christ that ever was on earth since the days of the Apostles" [69]—was not its merciless intolerance, for no one yet dreamed that tolerance was possible. It was the attempt to make the law of God prevail even in those matters of pecuniary gain and loss which mankind, to judge by its history, is disposed to regard more seriously than wounds and deaths. "No member [of the Christian body]," wrote Calvin in his *Institutes,* "holds his gifts to himself, or for his private use, but shares them among his fellow members, nor does he derive benefit save from those things which proceed from the common profit of the body as a whole. Thus the pious man owes to his brethren all that it is in his power to give." [70] It was natural that so remorseless an attempt to claim the totality of human interests for religion should not hesitate to engage even the economic appetites, before which the Churches of a later generation were to lower their arms. If Calvinism welcomed the world of business to its fold with an eagerness unknown before, it did so in the spirit of a conqueror organizing a new province, not of a suppliant arranging a compromise with a still powerful foe. A system of morals and a code of law lay ready to its hand in the Old Testament. Samuel and Agag, King of the Amalekites, Jonah and Nineveh, Ahab and Naboth, Elijah and the prophets of Baal, Micaiah the son of Imlah, the only true prophet of the Lord, and Jeroboam the son of Nebat, who made Israel to sin, worked on the tense imagination of the Calvinist as did Brutus and Cassius on the men of 1793. The first half-century of the Reformed Church at Geneva saw a prolonged effort to organize an economic order worthy of the Kingdom of Christ, in which the ministers played the part of Old Testament prophets to an Israel not wholly weaned from the fleshpots of Egypt.

Apart from its qualified indulgence to interest, Calvinism made few innovations in the details of social policy, and the contents of the program were thoroughly medieval. The novelty consisted in the religious zeal which was thrown into its application. The organ of administration before which offenders were brought was the Consistory, a mixed body of laymen and ministers. It censures harsh creditors, punishes usurers, engrossers and monopolists, reprimands or fines the merchant who defrauds his clients, the clothmaker whose stuff is an inch too narrow, the dealer who provides short measure of coal, the butcher who sells meat above the rates fixed by authority, the tailor who charges strangers excessive prices, the surgeon who demands an excessive fee for an operation.[71] In the Consistory the ministers appear to have carried all before them, and they are constantly pressing for greater stringency. From the election of Beza in place of Calvin in 1564 to his death in 1605, hardly a year passes without a new demand for legislation from the clergy, a new censure on economic unrighteousness, a new protest against one form or another of the ancient sin of avarice. At one moment, it is excessive indulgence to debtors which rouses their indignation; at another, the advance of prices and rents caused by the influx of distressed brethren from the persecutions in France; at a third, the multiplication of taverns and the excessive charges demanded by the sellers of wine. Throughout there is a prolonged warfare against the twin evils of extortionate interest and extortionate prices.

Credit was an issue of moment at Geneva, not merely

for the same reasons which made it a burning question everywhere to the small producer of the sixteenth century, but because, especially after the ruin of Lyons in the French wars of religion, the city was a financial center of some importance. It might be involved in war at any moment. In order to secure command of the necessary funds, it had borrowed heavily from Basle and Berne, and the Council used the capital to do exchange business and make advances, the rate of interest being fixed at 10, and later at 12, per cent. To the establishment of a bank the ministers, who had been consulted, agreed; against the profitable business of advancing money at high rates of interest to private persons they protested, especially when the loans were made to spendthrifts who used them to ruin themselves. When, ten years later, in 1580, the Council approved the project advanced by some company promoters of establishing a second bank in the city, the ministers led the opposition to it, pointed to the danger of covetousness as revealed by the moral corruption of financial cities such as Paris, Venice and Lyons, and succeeded in getting the proposal quashed. Naturally, however, the commoner issue was a more simple one. The capitalist who borrowed in order to invest and make a profit could take care of himself, and the ministers explained that they had no objection to those "qui baillent leur argent aux marchands pour emploier en marchandise." The crucial issue was that of the money-lender who makes advances "simplement à un qui aura besoin," and who thereby exploits the necessities of his poorer neighbors.[72]

Against monsters of this kind the ministers rage without ceasing. They denounce them from the pulpit in the name of the New Testament, in language drawn principally from the less temperate portions of the Old, as *larrons, brigands, loups et tigres,* who ought to be led out of the city and stoned to death. "The poor cry and the rich pocket their gains: but what they are heaping up for themselves is the wrath of God. . . . One has cried in the market-place, 'a curse on those who bring us dearth.' . . . The Lord has heard that cry . . . and yet we are asking the cause of the pestilence! . . . A cut-purse shall be punished, but the Lord declares by his prophet Amos . . . 'Famine is come upon my people of Israel, O ye who devour the poor.' The threats there uttered have been executed against his people." [73] They demand that for his second offense the usurer shall be excommunicated, or that, if such a punishment be thought too severe, he shall at least be required to testify his repentance publicly in church, before being admitted to the sacrament. They remind their fellow-citizens of the fate of Tyre and Sidon, and, momentarily despairing of controlling the money-lender directly, they propose to deprive him of his victims by removing the causes which create them. *Pour tarir les ruisseaux il faut escouper la source.* Men borrow because of "idleness, foolish extravagance, foolish sins, and law suits." Let censors be established at Geneva, as in Republican Rome, to inquire, among rich as well as among poor, how each household earns its livelihood, to see that all children of ten to twelve are taught some useful trade, to put down taverns and litigation, and to "bridle the insatiable avarice of those who are such wretches that they seek to enrich themselves by the necessities of their poor neighbors." [74]

The Venerable Company advanced their program, but they were not sanguine that it would be carried out, and they concluded it by expressing to the City Fathers the pious hope, not wholly free from irony, that "none of your honorable fellowship may be found spotted with such vices." Their apprehensions were justified. The Council of Geneva endured many things at the hands of its preachers, till, on the death of Beza, it brought them to heel. But there were limits to its patience, and it was in the field of business ethics that they were most quickly reached. It did not venture to question the right of the clergy to be heard on matters of commerce and finance. The pulpit was press and platform in one; ministers had the public behind them, and, conscious of their power, would in the last resort compel submission by threatening to resign *en masse*. Profuse in expressions of sympathy, its strategy was to let the cannon balls of Christian Socialism spend themselves on the yielding down of official procrastination, and its first reply was normally *qu'on y pense un peu*. To the clergy its inactivity was a new proof of complicity with Mammon, and they did not hesitate to declare their indignation from the pulpit. In 1574 Beza preached a sermon in which he accused members of the Council of having intelligence with speculators who had made a corner in wheat. Throughout 1577 the ministers were reproaching the Council with laxity in administration, and they finally denounced it as the real author of the rise in the prices of bread and wine. In 1579 they addressed to it a memorandum, setting out a new scheme of moral discipline and social reform.

The prosperous *bourgeoisie* who governed Geneva had no objection to discouraging extravagance in dress, or to exhorting the public to attend sermons and to send their children to catechism. But they heard denunciations of covetousness without enthusiasm, and on two matters they were obdurate. They refused to check, as the ministers concerned to lower prices had demanded, the export of wine, on the ground that it was needed in order to purchase imports of wheat; and, as was natural in a body of well-to-do creditors, they would make no concession to the complaint that debtors were subjected to a "double usury," since they were compelled to repay loans in an appreciating currency. Money fell as well as rose, they replied, and even the late M. Calvin, by whom the ordinance now criticized had been approved, had never pushed his scruples to such lengths. Naturally, the ministers were indignant at these evasions.

They informed the Council that large sums were being spent by speculators in holding up supplies of corn, and launched a campaign of sermons against avarice, with appropriate topical illustrations. Equally naturally, the Council retorted by accusing Beza of stirring up class hatred against the rich.[75]

The situation was aggravated by an individual scandal. One of the magistrates, who regarded Beza's remarks as a personal reflection, was rash enough to demand to be heard before the Council, with the result that he was found guilty, condemned to pay a fine, and compelled to forfeit fifty crowns which he had lent at 10 per cent. interest. Evidently, when matters were pushed to such lengths as this, no one, however respectable, could feel sure that he was safe. The Council and the ministers had already had words over the sphere of their respective functions, and were to fall out a year or two later over the administration of the local hospital. On this occasion the Council complained that the clergy were interfering with the magistrates' duties, and implied politely that they would be well advised to mind their own business.

So monstrous a suggestion—as though there were any human activity which was not the business of the Church!—evoked a counter-manifesto on the part of the ministers, in which the full doctrine of the earthly Jerusalem was set forth in all its majesty. They declined to express regret for having cited before the Consistory those who sold corn at extortionate prices, and for refusing the sacrament to one of them. Did not Solomon say, "Cursed is he who keeps his corn in time of scarcity"? To the charge of intemperate language Chauvet replied that the Council had better begin by burning the books of the Prophets, for he had done no more than follow the example set by Hosea. "If we should be silent," said Beza, "what would the people say? That they are dumb dogs. . . . As to the question of causing scandals, for the last two years there has been unceasing talk of usury, and, for all that, no more than three or four usurers have been punished. . . . It is notorious everywhere that the city is full of usurers, and that the ordinary rate is 10 per cent. or more." [76] The magistrates renewed their remonstrances. They had seen without a shudder an adulterer condemned to be hanged, and had mercifully commuted his sentence to scourging through the town, followed by ten years' imprisonment in chains. [77] But at the godly proposal to make capitalists die the death of Achan their humanity blenched. Besides, the punishment was not only cruel, but dangerous. In Geneva, "most men are debtors." If they are allowed to taste blood, who can say where their fury will end? Yet, such is the power of the spoken word, the magistrates did not venture on a blunt refusal, but gave scripture for scripture. They informed the ministers that they proposed to follow the example of David, who, when rebuked by Nathan, confessed his fault. Whether the ministers replied in the language of Nathan, we are not informed.

Recent political theory has been prolific in criticisms of the omnicompetent State. The principle on which the collectivism of Geneva rested may be described as that of the omnicompetent Church. [78] The religious community formed a closely organized society, which, while using the secular authorities as police officers to enforce its mandates, not only instructed them as to the policy to be pursued, but was itself a kind of State, prescribing by its own legislation the standard of conduct to be observed by its members, putting down offences against public order and public morals, providing for the education of youth and for the relief of the poor. The peculiar relations between the ecclesiastical and secular authorities, which for a short time made the system possible at Geneva, could not exist to the same degree when Calvinism was the creed, not of a single city, but of a mi-

nority in a national State organized on principles quite different from its own.   Unless the State itself were captured,
rebellion, civil war or the abandonment of the pretension
to control society was the inevitable consequence.   But the
last result was long delayed.   In the sixteenth century, whatever the political conditions, the claim of the Calvinist
Churches is everywhere to exercise a collective responsibility
for the moral conduct of their members in all the various
relations of life, and to do so, not least, in the sphere of
economic transactions, which offer peculiarly insidious
temptations to a lapse into immorality.

The mantle of Calvin's system fell earliest upon the Reformed Churches of France.   At their first Synod, held in
1559 at Paris, where a scheme of discipline was adopted,
certain difficult matters of economic casuistry were discussed, and similar questions continued to receive attention
at subsequent Synods for the next half-century, until, as
the historian of French Calvinism remarks, "they began to
lax the reins, yielding too much to the iniquity of the
time." [79]   Once it is admitted that membership of the Church
involves compliance with a standard of economic morality
which the Church must enforce, the problems of interpretation which arise are innumerable, and the religious community finds itself committed to developing something like
a system of case law, by the application of its general principles to a succession of varying situations.   The elaboration
of such a system was undertaken; but it was limited in the
sixteenth century both by the comparative simplicity of the
economic structure, and by the fact that the Synods, except
at Geneva, being concerned not to reform society, but merely
to repress the grosser kinds of scandal, dealt only with
matters on which specific guidance was demanded by the
Churches.

Even so, however, the riddles to be solved were not a
few.   What is to be the attitude of the Churches towards

those who have grown rich on ill-gotten wealth? May pirates and fraudulent tradesmen be admitted to the Lord's Supper? May the brethren trade with such persons, or do they share their sin if they buy their goods? The law of the State allows moderate interest: what is to be the attitude of the Church? What is to be done to prevent craftsmen cheating the consumer with shoddy wares, and tradesmen oppressing him with extortionate profits? Are lotteries permissible? Is it legitimate to invest at interest monies bequeathed for the benefit of the poor? The answers which the French Synods made to such questions show the persistence of the idea that the transactions of business are the province of the Church, combined with a natural desire to avoid an impracticable rigor. All persons who have wrung wealth unjustly from others must make restitution before they be admitted to communion, but their goods may be bought by the faithful, provided that the sale is public and approved by the civil authorities. Makers of fraudulent wares are to be censured, and tradesmen are to seek only "indifferent gain." On the question of usury, the same division of opinion is visible in the French Reformed Church as existed at the same time in England and Holland, and Calvin's advice on the subject was requested. The stricter school would not hear of confining the prohibition of usury to "excessive and scandalous" exactions, or of raising money for the poor by interest on capital. In France, however, as elsewhere, the day for these heroic rigors had passed, and the common-sense view prevailed. The brethren were required to demand no more than the law allowed and than was consistent with charity. Within these limits interest was not to be condemned.[80]

Of the treatment of questions of this order by English Puritanism something is said in a subsequent chapter. In Scotland the views of the reformers as to economic ethics did not differ in substance from those of the Church before the Reformation, and the Scottish Book of Discipline denounced covetousness with the same vehemence as did the "accursed Popery" which it had overthrown. Gentlemen are exhorted to be content with their rents, and the Churches are required to make provision for the poor. "Oppression of the poor by exactions," it is declared, "[and] deceiving of them in buying or selling by wrong mete or measure . . . do properly appertain to the Church of God, to punish the same as God's word commandeth." [81] The interpretation given to these offences is shown by the punishment of a usurer and of a defaulting debtor before the Kirk Sessions of St. Andrews.[82] The relief of the poor was in 1579 made the statutory duty of ecclesiastical authorities in Scotland, seven years after it had in England been finally transferred to the State. The arrangement under which in rural districts it reposed down to 1846 on the shoulders of ministers, elders and deacons, was a survival from an age in which the real State in Scotland had been represented, not by Parliament or Council, but by the Church of Knox.

Of English-speaking communities, that in which the social discipline of the Calvinist Church-State was carried to the furthest extreme was the Puritan theocracy of New England. Its practice had more affinity with the iron rule of Calvin's Geneva than with the individualistic tendencies of contemporary English Puritanism. In that happy, bishopless Eden, where men desired only to worship God "according to the simplicitie of the gospel and to be ruled by the laws of God's word," [83] not only were "tobacco and immodest fashions and costly apparel," and "that vain custom of drinking one to another," forbidden to true professors, but the Fathers adopted towards that "notorious evil . . . whereby most men walked in all their commerce—to buy as cheap and sell as dear as they can," [84] an attitude which possibly would not be wholly congenial to their more businesslike descendants. At an early date in the history of Mas-

sachusetts a minister had called attention to the recru-
descence of the old Adam—"profit being the chief aim and
not the propagation of religion"—and Governor Bradford,
observing uneasily how men grew "in their outward estates," remarked that the increase in material prosperity
"will be the ruin of New England, at least of the Churches
of God there." [85]    Sometimes Providence smote the ex-
ploiter.  The immigrant who organized the first American
Trust—he owned the only milch cow on board and sold
the milk at 2d. a quart—"being after at a sermon wherein
oppression was complained of . . . fell distracted." [86] Those
who escaped the judgment of Heaven had to face the civil
authorities and the Church, which, in the infancy of the
colony, were the same thing.

Naturally the authorities regulated prices, limited the rate
of interest, fixed a maximum wage, and whipped incor-
rigible idlers; for these things had been done even in the
house of bondage from which they fled.  What was more
distinctive of the children of light was their attempt to
apply the same wholesome discipline to the elusive category
of business profits.  The price of cattle, the Massachusetts
authorities decreed, was to be determined, not by the needs
of the buyer, but so as to yield no more than a reasonable
return to the seller.[87]  Against those who charged more,
their wrath was that of Moses descending to find the chosen
people worshipping a golden calf.  What little emotion they
had to spare from their rage against religious freedom, they
turned against economic license.  Roger Williams touched
a real affinity when, in his moving plea for tolerance, he
argued that, though extortion was an evil, it was an evil
the treatment of which should be left to the discretion of
the civil authorities.[88]

Consider the case of Mr. Robert Keane.  His offence, by
general consent, was black.  He kept a shop in Boston, in

which he took "in some . . . above 6*d*. in the shilling profit; in some above 8*d.;* and in some small things above two for one"; and this, though he was "an ancient professor of the gospel, a man of eminent parts, wealthy and having but one child, having come over for conscience' sake and for the advancement of the gospel." The scandal was terrible. Profiteers were unpopular—"the cry of the country was great against oppression"—and the grave elders reflected that a reputation for greed would injure the infant community, lying as it did "under the curious observation of all Churches and civil States in the world." In spite of all, the magistrates were disposed to be lenient. There was no positive law in force limiting profits; it was not easy to determine what profits were fair; the sin of charging what the market could stand was not peculiar to Mr. Keane; and, after all, the law of God required no more than double restitution. So they treated him mercifully, and fined him only £200.

Here, if he had been wise, Mr Keane would have let the matter drop. But, like some others in a similar position, he damned himself irretrievably by his excuses. Summoned before the church of Boston, he first of all "did with tears acknowledge and bewail his covetous and corrupt heart," and then was rash enough to venture on an explanation, in which he argued that the tradesman must live, and how could he live, if he might not make up for a loss on one article by additional profit on another? Here was a text on which no faithful pastor could refrain from enlarging. The minister of Boston pounced on the opportunity, and took occasion "in his public exercise the next lecture day to lay open the error of such false principles, and to give some rules of direction in the case. Some false principles were these:—

"1. That a man might sell as dear as he can, and buy as cheap as he can.

"2. If a man lose by casualty of sea, etc., in some of his commodities, he may raise the price of the rest.

"3. That he may sell as he bought, though he paid too dear, and though the commodity be fallen, etc.

"4. That, as a man may take the advantage of his own skill or ability, so he may of another's ignorance or necessity.

"5. Where one gives time for payment, he is to take like recompence of one as of another."

The rules for trading were not less explicit:—

"1. A man may not sell above the current price, i.e., such a price as is usual in the time and place, and as another (who knows the worth of the commodity) would give for it if he had occasion to use it; as that is called current money which every man will take, etc.

"2. When a man loseth in his commodity for want of skill, etc., he must look at it as his own fault or cross, and therefore must not lay it upon another.

"3. Where a man loseth by casualty of sea, etc., it is a loss cast upon himself by Providence, and he may not ease himself of it by casting it upon another; for so a man should seem to provide against all providences, etc., that he should never lose; but where there is a scarcity of the commodity, there men may raise their price; for now it is a hand of God upon the commodity, and not the person.

"4. A man may not ask any more for his commodity than his selling price, as Ephron to Abraham: the land is worth thus much."

It is unfortunate that the example of Ephron was not remembered in the case of transactions affecting the lands of Indians, to which it might have appeared peculiarly appropriate. In negotiating with these children of the devil, however, the saints of God considered the dealings of Israel with Gibeon a more appropriate precedent.

The sermon was followed by an animated debate within

the church. It was moved, amid quotations from 1 Cor. v.
11, that Mr. Keane should be excommunicated. That he
might be excommunicated, if he were a covetous person
within the meaning of the text, was doubted as little as
that he had recently given a pitiable exhibition of covetousness. The question was only whether he had erred through
ignorance or careless, or whether he had acted "against his
conscience or the very light of nature"—whether, in short,
his sin was accidental or a trade. In the end he escaped
with his fine and admonition.[89]

If the only Christian documents which survived were the
New Testament and the records of the Calvinist Churches
in the age of the Reformation, to suggest a connection between them more intimate than a coincidence of phraseology
would appear, in all probability, a daring extravagance.
Legalistic, mechanical, without imagination or compassion,
the work of a jurist and organizer of genius, Calvin's system was more Roman than Christian, and more Jewish than
either. That it should be as much more tyrannical than
the medieval Church, as the Jacobin Club was than the
*ancien régime,* was inevitable. Its meshes were finer, its
zeal and its efficiency greater. And its enemies were not
merely actions and writings, but thoughts.

The tyranny with which it is reproached by posterity
would have been regarded by its champions as a compli-
ment. In the struggle between liberty and authority, Calvinism sacrificed liberty, not with reluctance, but with enthusiasm. For the Calvinist Church was an army march-
ing back to Canaan, under orders delivered once for all
from Sinai, and the aim of its leaders was the conquest of
the Promised Land, not the consolation of stragglers or the
encouragement of laggards. In war the classical expedient
is a dictatorship. The dictatorship of the ministry appeared as inevitable to the whole-hearted Calvinist as the
Committee of Public Safety to the men of 1793, or the

dictatorship of the proletariat to an enthusiastic Bolshevik. If it reached its zenith where Calvin's discipline was accepted without Calvin's culture and intellectual range, in the orgies of devil worship with which a Cotton and an Endicott shocked at last even the savage superstition of New England, that result was only to be expected.

The best that can be said of the social theory and practice of early Calvinism is that they were consistent. Most tyrannies have contented themselves with tormenting the poor. Calvinism had little pity for poverty; but it distrusted wealth, as it distrusted all influences that distract the aim or relax the fibers of the soul, and, in the first flush of its youthful austerity, it did its best to make life unbearable for the rich. Before the Paradise of earthly comfort it hung a flaming brand, waved by the implacable shades of Moses and Aaron.[90]

# CHAPTER III

## THE CHURCH OF ENGLAND

"If any man be so addicted to his private, that he neglect the common, state, he is void of the sense of piety, and wisheth peace and happiness to himself in vain. For, whoever he be, he must live in the body of the Commonwealth and in the body of the Church."

LAUD, *Sermon before His Majesty,* June 19, 1621.

# CHAPTER III

THE ecclesiastical and political controversies which descend from the sixteenth century have thrust into oblivion all issues of less perennial interest. But the discussions which were motived by changes in the texture of society and the relations of classes were keen and continuous, nor was their result without significance for the future. In England, as on the Continent, the new economic realities came into sharp collision with the social theory inherited from the Middle Ages. The result was a re-assertion of the traditional doctrines with an almost tragic intensity of emotion, their gradual retreat before the advance of new conceptions, both of economic organization and of the province of religion, and their final decline from a militant creed into a kind of pious antiquarianism. They lingered, venerable ghosts, on the lips of churchmen down to the Civil War. Then the storm blew and they flickered out.

Medieval England had lain on the outer edge of economic civilization, remote from the great highways of commerce and the bustling financial centers of Italy and Germany. With the commercial revolution which followed the Discoveries, a new age began. After the first outburst of curiosity, interest in explorations which yielded no immediate return of treasure died down. It was not till more than half a century later, when the silver of the New World was dazzling all Europe, that Englishmen reflected that it might conceivably have been lodged in the Tower instead of at Seville, and that talk of competition for America and the East began in earnest.

In the meantime, however, every other aspect of English economic life was in process of swift transformation. Foreign trade increased largely in the first half of the sixteenth century, and, as manufactures developed, cloth displaced wool as the principal export. With the growth of commerce went the growth of the financial organization on which commerce depends, and English capital poured into the growing London money-market, which had previously been dominated by Italian bankers. At home, with the expansion of internal trade which followed the Tudor peace, opportunities of speculation were increased, and a new class of middlemen arose to exploit them. In industry, the rising interest was that of the commercial capitalist, bent on securing the freedom to grow to what stature he could, and produce by what methods he pleased. Hampered by the defensive machinery of the gilds, with their corporate discipline, their organized torpor restricting individual enterprise, and their rough equalitarianism, either he quietly evaded gild regulations by withdrawing from the corporate towns, within which alone the pressure of economic conformity could be made effective, or he accepted the gild organization, captured its government, and by means of it developed a system under which the craftsman, even if nominally a master, was in effect the servant of an employer. In agriculture, the customary organization of the village was being sapped from below and battered down from above. For a prosperous peasantry, who had commuted the labor services that were still the rule in France and Germany, were rearranging their strips by exchange or agreement, and lords, no longer petty sovereigns, but astute business men, were leasing their demesnes to capitalist farmers, quick to grasp the profits to be won by sheep-grazing, and eager to clear away the network of communal restrictions which impeded its extension. Into commerce, industry and agriculture alike, the revolution in prices, gradual for the first third of the century, but after 1540 a mill race, injected a virus of hitherto unsuspected potency, at once a stimulant to feverish enterprise and an acid dissolving all customary relationships.

It was a society in rapid motion, swayed by new ambitions and haunted by new terrors, in which both success and failure had changed their meaning. Except in the turbulent north, the aim of the great landowner was no longer to hold at his call an army of retainers, but to exploit his estates as a judicious investment. The prosperous merchant, once content to win a position of dignity and power in fraternity or town, now flung himself into the task of carving his way to solitary preëminence, unaided by the artificial protection of gild or city. To the immemorial poverty of peasant and craftsman, pitting, under the ever-present threat of famine, their pigmy forces against an implacable nature, was added the haunting insecurity of a growing, though still small, proletariat, detached from their narrow niche in village or borough, the sport of social forces which they could neither understand, nor arrest, nor control.

## I. THE LAND QUESTION

The England of the Reformation, to which posterity turns as a source of high debates on church government and doctrine, was to contemporaries a cauldron seething with economic unrest and social passions. But the material on which agitation fed had been accumulating for three generations, and of the grievances which exploded in the middle of the century, with the exception of the depreciation of the currency, there was not one—neither enclosures and pasture farming, nor usury, nor the malpractices of gilds, nor the rise in prices, nor the oppression of craftsmen by merchants, nor the extortions of the engrosser—which had not evoked popular protests, been denounced by publicists, and produced legislation and administrative action, long before the Reformation Parliament met. The floods were already running high, when the religious revolution swelled them with a torrent of bitter, if bracing, waters. Its effect on the social situation was twofold. Since it produced a sweeping redistribution of wealth, carried out by an unscrupulous minority using the weapons of violence, intimidation and fraud, and succeeded by an orgy of interested misgovernment on the part of its principal beneficiaries, it aggravated every problem, and gave a new turn to the screw which was squeezing peasant and craftsman. Since it released a torrent of writing on questions not only of religion, but of social organization, it caused the criticisms passed on the changes of the past half-century to be brought to a head in a sweeping indictment of the new economic forces and an eloquent restatement of the traditional theory of social obligations. The center of both was the land question. For it was agrarian plunder which principally stirred the cupidity of the age, and agrarian grievances which were the most important ground of social agitation.

The land question had been a serious matter for the greater part of a century before the Reformation. The first detailed account of enclosure had been written by a chantry priest in Warwickshire, soon after 1460.[1] Then had come the legislation of 1489, 1515 and 1516, Wolsey's Royal Commission in 1517, and more legislation in 1534.[2] Throughout, a steady stream of criticism had flowed from men of the Renaissance, like More, Starkey and a host of less well-known writers, dismayed at the advance of social anarchy, and sanguine of the miracles to be performed by a Prince who would take counsel of philosophers.

If, however, the problem was acute long before the confiscation of the monastic estates, its aggravation by the fury of spoliation let loose by Henry and Cromwell is not open to serious question. It is a mistake, no doubt, to see the last days of monasticism through rose-colored spectacles. The monks, after all, were business men, and the lay agents whom they often employed to manage their property naturally conformed to the agricultural practice of the world around them. In Germany revolts were nowhere more frequent or more bitter than on the estates of ecclesiastical land-owners.[3] In England a glance at the proceedings of the Courts of Star Chamber and Requests is enough to show that holy men reclaimed villeins, turned copy-holders into tenants at will, and, as More complained, converted arable land to pasture.[4]

In reality, the supposition of unnatural virtue on the part of the monks, or of more than ordinary harshness on the part of the new proprietors, is not needed in order to explain the part which the rapid transference of great masses of property played in augmenting rural distress. The worst side of all such sudden and sweeping redistributions is that the individual is more or less at the mercy of the market, and can hardly help taking his pound of flesh. Estates with a capital value (in terms of modern money) of £15,000,000 to £20,000,000 changed hands.[5] To the abbey lands which came into the market after 1536 were added those of the gilds and chantries in 1547. The financial necessities of the Crown were too pressing to allow of its retaining them in its own possession and drawing the rents; nor, in any case, would that have been the course dictated by prudence to a Government which required a party to carry through a revolution. What it did, therefore, was to alienate most of the land almost immediately, and to spend the capital as income. For a decade there was a mania of land speculation. Much of the property was bought by needy courtiers, at a ridiculously low figure. Much of it passed to sharp business men, who brought to bear on its management the methods learned in the financial school of the City; the largest single grantee was Sir Richard Gres-

ham.  Much was acquired by middlemen, who bought scattered parcels of land, held them for the rise, and disposed
of them piecemeal when they got a good offer; in London,
groups of tradesmen—cloth-workers, leather-sellers, merchant tailors, brewers, tallow-chandlers—formed actual
syndicates to exploit the market.  Rack-renting, evictions,
and the conversions of arable to pasture were the natural
result, for surveyors wrote up values at each transfer, and,
unless the last purchaser squeezed his tenants, the transaction would not pay.[6]

Why, after all, should a landlord be more squeamish than
the Crown?  "Do ye not know," said the grantee of one
of the Sussex manors of the monastery of Sion, in answer
to some peasants who protested at the seizure of their commons, "that the King's Grace hath put down all the houses
of monks, friars and nuns?  Therefore now is the time
come that we gentlemen will pull down the houses of such
poor knaves as ye be."[7]  Such arguments, if inconsequent,
were too convenient not to be common.  The protests of contemporaries receive detailed confirmation from the bitter
struggles which can be traced between the peasantry and
some of the new landlords—the Herberts, who enclosed a
whole village to make the park at Washerne, in which, according to tradition, the gentle Sidney was to write his *Ar-
cadia,* the St. Johns at Abbot's Ripton, and Sir John Yorke,
third in the line of speculators in the lands of Whitby Abbey, whose tenants found their rents raised from £29 to £64
a year, and for nearly twenty years were besieging the Government with petitions for redress.[8]  The legend, still repeated late in the seventeenth century, that the grantees of
monastic estates died out in three generations, though un-
veracious, is not surprising.  The wish was father to the
thought.

It was an age in which the popular hatred of the en-
closer and the engrosser found a natural ally in religious sen-

timent, schooled, as it was, in a tradition which had taught
that the greed of gain was a deadly sin, and that the plea
of economic self-interest did not mitigate the verdict, but
aggravated the offence.   In England, as on the Continent,
doctrinal radicalism marched hand in hand with social conservatism.   The most scathing attack on social disorders
came, not from the partisans of the old religion, but from
divines on the left wing of the Protestant party, who saw
in economic individualism but another expression of the
laxity and license which had degraded the purity of religion, and who understood by reformation a return to the
moral austerity of the primitive Church, no less than to its
government and doctrine.   The touching words [9] in which
the leader of the Pilgrimage of Grace painted the social effects of the dissolution of the Yorkshire monasteries were
mild compared with the denunciations launched ten years
later by Latimer, Crowley, Lever, Becon and Ponet.

Their passion was natural.   What Aske saw in the green
tree, they saw in the dry, and their horror at the plunge into
social immorality was sharpened by the bitterness of disappointed hopes.   It was all to have been so different!   The
movement which produced the Reformation was a Janus,
not with two, but with several, faces, and among them
had been one which looked wistfully for a political and
social regeneration as the fruit of the regeneration of religion.[10]   In England, as in Germany and Switzerland, men
had dreamed of a Reformation which would reform the
State and society, as well as the Church.   The purification,
not merely of doctrine, but of morals, the encouragement of
learning, the diffusion of education, the relief of poverty,
by the stirring into life of a mass of sleeping endowments,
a spiritual and social revival inspired by the revival of the
faith of the Gospel—such, not without judicious encouragement from a Government alert to play on public opinion,

was the vision which had floated before the eyes of the humanitarian and the idealist.

It did not vanish without a struggle. At the very height of the economic crisis, Bucer, the tutor of Edward VI, and Professor of Divinity at Cambridge, stated the social program of a Christian renaissance in the manual of Christian politics which he drafted in order to explain to his pupil how the Kingdom of Christ might be established by a Christian prince. Its outlines were sharpened, and its details elaborated, with all the remorseless precision of a disciple of Calvin. Willful idlers are to be excommunicated by the Church and punished by the State. The Government, a pious mercantilist, is to revive the woollen industry, to introduce the linen industry, to insist on pasture being put under the plow. It is to take a high line with the commercial classes. For, though trade in itself is honorable, most traders are rogues—indeed "next to the sham priests, no class of men is more pestilential to the Commonwealth"; their works are usury, monopolies, and the bribery of Governments to overlook both. Fortunately, the remedies are simple. The State must fix just prices—"a very necessary but an easy matter." Only "pious persons, devoted to the Commonwealth more than to their own interests," are to be allowed to engage in trade at all. In every village and town a school is to be established under a master eminent for piety and wisdom. "Christian princes must above all things strive that men of virtue may abound, and live to the glory of God. . . . Neither the Church of Christ, nor a Christian Commonwealth, ought to tolerate such as prefer private gain to the public weal, or seek it to the hurt of their neighbors." [11]

The Christian prince strove, but not, poor child, as those that prevail. The classes whose backing was needed to make the Reformation a political success had sold their support on terms which made it inevitable that it should be a social disaster. The upstart aristocracy of the future had their teeth in the carcass, and, having tasted blood, they were not to be whipped off by a sermon. The Government of Edward VI, like all Tudor Governments, made its experiment in fixing just prices. What the astute Gresham, its financial adviser, thought of restricting commerce to persons of piety, we do not know, but can guess. As for the schools, what it did for them Mr. Leach has told us. It swept them away wholesale in order to distribute their endowments among courtiers. There were probably more schools in proportion to the population at the end of the fifteenth century than there were in the middle of the nineteenth. "These endowments were confiscated by the State, and many still line the pockets of the descendants of the statesmen of the day." [12] "King Edward VI's Grammar Schools" are the schools which King Edward VI did not destroy.

The disillusionment was crushing. Was it surprising that the reformers should ask what had become of the devout imaginations of social righteousness, which were to have been realized as the result of a godly reformation? The end of Popery, the curtailment of ecclesiastical privileges, six new bishoprics, lectureships in Greek and Latin in place of the disloyal subject of the canon law, the reform of doctrine and ritual—side by side with these good things had come some less edifying changes, the ruin of much education, the cessation of much charity, a raid on corporate property which provoked protests even in the House of Commons, [13] and for ten years a sinister hum, as of the floating of an immense land syndicate, with favorable terms for all sufficiently rich, or influential, or mean, to get in on the ground floor. The men who had invested in the Reformation when it was still a gambling stock naturally nursed the security, and denounced the revolting peasants as communists, with the mystical reverence for the rights of property which is characteristic in all ages of the *nou-*

*veaux riches.*[14]    The men whose religion was not money said what they thought of the business in pamphlets and sermons, which left respectable congregations spluttering with fury.

Crowley pilloried lease-mongers and usurers, wrote that the sick begged in the street because rich men had seized the endowments of hospitals, and did not conceal his sympathy with the peasants who rose under Ket.[15]    Becon told the gentry, eloquent on the vices of abbey-lubbers, that the only difference between them and the monks was that they were more greedy and more useless, more harsh in wringing the last penny from the tenants, more selfish in spending the whole income on themselves, more pitiless to the poor.[16]    "In suppressing of abbies, cloisters, colleges and chantries," preached Lever in St. Paul's, "the intent of the King's Majesty that dead is, was, and of this our king now is, very godly, and the purpose, or else the pretence, of other wondrous goodly: that thereby such abundance of goods as was superstitiously spent upon vain ceremonies, or voluptuously upon idle bellies, might come to the king's hands to bear his great charges, necessarily bestowed in the common wealth, or partly unto other men's hands, for the better relief of the poor, the maintenance of learning, and the setting forth of God's word.    Howbeit, covetous officers have so used this matter, that even those goods which did serve to the relief of the poor, the maintenance of learning, and to comfortable necessary hospitality in the common wealth, be now turned to maintain worldly, wicked, covetous ambition. . . . You which have gotten these goods into your own hands, to turn them from evil to worse, and other goods more from good unto evil, be ye sure it is even you that have offended God, beguiled the king, robbed the rich, spoiled the poor, and brought a common wealth into a common misery." [17]

This was plain speaking indeed.    Known to their enemies

as the "Commonwealth men" from their advocacy of social reconstruction, the group of which Latimer was the prophet and Hales the man of action naturally incurred the charge of stirring up class-hatred, which is normally brought against all who call attention to its causes. The result of their activity was the appointment of a Royal Commission to inquire into offences against the Acts forbidding the conversion of arable to pasture, the introduction of legislation requiring the maintenance of tillage and rebuilding of cottages, and a proclamation pardoning persons who had taken the law into their own hands by pulling down hedges. The gentry were furious. Paget, the secretary to the Council, who was quite ready for a reign of terror, provided that the gentlemen began it, prophesied gloomily that the German Peasants' War was to be reënacted in England; the Council, most of whose members held abbey lands, was sullen; and Warwick, the personification of the predatory property of the day, attacked Hales fiercely for carrying out, as chairman of the Midland committee of the Depopulation Commission, the duties laid upon him by the Government. "Sir," wrote a plaintiff gentleman to Cecil, "be plain with my Lord's Grace, that under the pretense of simplicity and poverty there may [not] rest much mischief. So do I fear there doth in these men called Common Wealths and their adherents. To declare unto you the state of the gentlemen (I mean as well the greatest as the lowest), I assure you they are in such doubt, that almost they dare touch none of them [i.e., the peasants], not for that they are afraid of them, but for that some of them have been sent up and come away without punishment, and that Common Wealth called Latimer hath gotten the pardon of others." [18]

The "Common Wealth called Latimer" was unrepentant. Combining gifts of humor and invective which are not very common among bishops, his fury at oppression did

not prevent him from greeting the Devil with a burst of
uproarious laughter, as of a satyrical gargoyle carved to
make the sinner ridiculous in this world before he is
damned in the next.  So he was delighted when he provoked one of his audience into the exclamation, "Mary, a
seditious fellow!" used the episode as comic relief in his
next sermon,[19] and then, suddenly serious, redoubled, his
denunciations of step-lords and rent-raisers.  Had not the
doom of the covetous been pronounced by Christ Himself?

> You thoughte that I woulde not requyre
> The bloode of all suche at your hande,
> But be you sure, eternall fyre
> Is redy for eche hell fyrebrande.
> Both for the housynge and the lande
> That you have taken from the pore
> Ye shall in hell dwell evermore.[20]

On the technicalities of the Tudor land question the
authors of such outbursts spoke without authority, and,
thanks to Mr. Leadam and Professor Gay, modern research
has found no difficulty in correcting the perspective of their
story.  At once incurious and ill-informed as to the large
impersonal causes which were hurrying forward the reorganization of agriculture on a commercial basis, what
shocked them was not only the material misery of their age,
but its repudiation of the principles by which alone, as it
seemed, human society is distinguished from a pack of
wolves.  Their enemy was not merely the Northumberlands
or Herberts, but an idea, and they sprang to the attack, less
of spoliation or tyranny, than of a creed which was the
parent of both.  That creed was that the individual is
absolute master of his own, and, within the limits set by
positive law, may exploit it with a single eye to his pecuniary
advantage, unrestrained by any obligation to postpone his
own profit to the well-being of his neighbors, or to give

account of his actions to a higher authority. It was, in short, the theory of property which was later to be accepted by all civilized communities.

The question of the respective rights of lord and peasant had never, at least within recent centuries, arisen in so acute a form, for, as long as the customary tenants were part of the stock of the manor, it was obviously to the interest of the lord to bind them to the soil. Now all that had been changed, at any rate in the south and midlands, by the expansion of the woollen industry and the devaluation of money. Chevage and merchet had gone; forced labor, if it had not gone, was fast going. The psychology of landowning had been revolutionized, and for two generations the sharp landlord, instead of using his seigneurial right to fine or arrest run-aways from the villein nest, had been hunting for flaws in titles, screwing up admission fines, twisting manorial customs, and, when he dared, turning copyholds into leases. The official opposition to depopulation, which had begun in 1489 and was to last almost till 1640, infuriated him, as an intolerable interference with the rights of property. In their attacks on the restraints imposed by village custom from below and by the Crown from above, in their illegal defiance of the statutes forbidding depopulation, and in their fierce resistance to the attempts of Wolsey and Somerset to restore the old order, the interests which were making the agrarian revolution were watering the seeds of that individualistic conception of ownership which was to carry all before it after the Civil War. With such a doctrine, since it denied both the existence and the necessity of a moral title, it was not easy for any religion less pliant than that of the eighteenth century to make a truce. Once accepted, it was to silence the preaching of all social duties save that of submission. If property be an unconditional right, emphasis on its obligations is little more than the graceful parade of a

flattering, but innocuous, metaphor. For, whether the obligations are fulfilled or neglected, the right continues unchallenged and indefeasible.

A religious theory of society necessarily regards with
suspicion all doctrines which claim a large space for the unfettered play of economic self-interest. To the latter the
end of activity is the satisfaction of desires, to the former
the felicity of man consists in the discharge of obligations
imposed by God. Viewing the social order as the imperfect reflection of a divine plan, it naturally attaches a high
value to the arts by which nature is harnessed to the service
of mankind. But, more concerned with ends than with
means, it regards temporal goods as at best instrumental
to a spiritual purpose, and its standpoint is that of Bacon,
when he spoke of the progress of knowledge as being
sought for "the glory of the Creator and the relief of
man's estate." To a temper nurtured on such ideas, the new
agrarian *régime,* with its sacrifice of the village—a fellowship of mutual aid, a partnership of service and protection,
"a little commonwealth"—to the pecuniary interests of a
great proprietor, who made a desert where men had worked
and prayed, seemed a defiance, not only of man, but of God.
It was the work of "men that live as thoughe there were no
God at all, men that would have all in their owne handes,
men that would leave nothyng for others, men that would
be alone on the earth, men that bee never satisfied." [21] Its
essence was an attempt to extend legal rights, while repudiating legal and quasi-legal obligations. It was against
this new idolatry of irresponsible ownership, a growing,
but not yet triumphant, creed, that the divines of the Reformation called down fire from heaven.

Their doctrine was derived from the conception of property, of which the most elaborate formulation had been made
by the Schoolmen, and which, while justifying it on grounds
of experience and expediency, insisted that its use was lim-

ited at every turn by the rights of the community and the obligations of charity. Its practical application was an idealized version of the feudal order, which was vanishing before the advance of more business-like and impersonal forms of land-ownership, and which, once an engine of exploitation, was now hailed as a bulwark to protect the weak against the downward thrust of competition. Society is a hierarchy of rights and duties. Law exists to enforce the second, as much as to protect the first. Property is not a mere aggregate of economic privileges, but a responsible office. Its *raison d'être* is not only income, but service. It is to secure its owner such means, and no more than such means, as may enable him to perform those duties, whether labor on the land, or labor in government, which are involved in the particular status which he holds in the system. He who seeks more robs his superiors, or his dependents, or both. He who exploits his property with a single eye to its economic possibilities at once perverts its very essence and destroys his own moral title, for he has "every man's living and does no man's duty." [22]

The owner is a trustee, whose rights are derived from the function which he performs and should lapse if he repudiates it. They are limited by his duty to the State; they are limited no less by the rights of his tenants against him. Just as the peasant may not cultivate his land in the way which he may think most profitable to himself, but is bound by the law of the village to grow the crops which the village needs and to throw his strips open after harvest to his neighbors's beasts, so the lord is required both by custom and by statute to forego the anti-social profits to be won by methods of agriculture which injure his neighbors and weaken the State. He may not raise his rent or demand increased fines, for the function of the peasant, though different, is not less essential than his own. He is, in short not a *rentier,* but an officer, and it is for the Church to re buke him when he sacrifices the duties of his charge to the greed for personal gain. "We heartily pray thee to send thy holy spirit into the hearts of them that possess the grounds, pastures, and dwelling-places of the earth, that they, remembering themselves to be thy tenants, may not rack and stretch out the rents of their houses and lands, nor yet take unreasonable fines and incomes, after the manner of covetous worldlings . . . but so behave themselves in letting out their tenements, lands and pastures, that after this life they may be received into everlasting dwelling places." [23] Thus, while the covetous worldlings disposed the goods of this transitory life to their liking, did a pious monarch consider their eternal welfare in the Book of Private Prayer issued in 1553.

## II. RELIGIOUS THEORY AND SOCIAL POLICY

If a philosophy of society is to be effective, it must be as mobile and realistic as the forces which it would control. The weakness of an attitude which met the onset of insurgent economic interests with a generalized appeal to traditional morality and an idealization of the past was only too obvious. Shocked, confused, thrown on to a helpless, if courageous and eloquent, defensive by changes even in the slowly moving world of agriculture, medieval social theory, to which the most representative minds of the English Church still clung, found itself swept off its feet after the middle of the century by the swift rise of a commercial civilization, in which all traditional landmarks seemed one by one to be submerged. The issue over which the struggle between the new economic movements of the age and the scheme of economic ethics expounded by churchmen was most definitely joined, and continued longest, was not, as the modern reader might be disposed to expect, that of wages, but that of credit, money-lending and prices. The center of the controversy—the mystery of iniquity in which a host of minor scandals were conveniently, if inaccurately, epitomized—was the problem which contemporaries described by the word usury.

"Treasure doth then advance greatness," wrote Bacon, in words characteristic of the social ideal of the age," when the wealth of the subject be rather in many hands than few." [24]   In spite of the growing concentration of property, Tudor England was still, to use a convenient modern phase, a Distributive State.  It was a community in which the ownership of land, and of the simple tools used in most industries, was not the badge of a class, but the attribute of a society, and in which the typical worker was a peasant farmer, a tradesman, or a small master.  In this world of small property-owners, of whose independence and prosperity English publicists boasted, in contrast with the "housed beggars" of France and Germany, the wage-earners were a minority scattered in the interstices of village and borough, and, being normally themselves the sons of peasants, with the prospect of stepping into a holding of their own, or, at worst, the chance of squatting on the waste, were often in a strong position *vis-à-vis* their employers.  The special economic *malaise* of an age is naturally the obverse of its special qualities.  Except in certain branches of the textile industry, the grievance which supplied fuel to social agitation, which evoked programs of social reform, and which prompted both legislation and administrative activity, sprang, not from the exploitation of a wage-earning proletariat by its employers, but from the relation of the producer to the landlord of whom he held, the dealer with whom he bought and sold, and the local capitalist, often the dealer in another guise, to whom he ran into debt.  The farmer must borrow money when the season is bad, or merely to finance the interval between sowing and harvest.  The craftsman must buy raw materials on credit and get advances before his wares are sold. The young tradesman must scrape together a little capital. before he can set up shop. Even the cottager, who buys grain at the local market, must constantly ask the seller to "give day." Almost every one, therefore, at one time or another, has need of the money-lender. And the lender is often a monopolist—"a money master," a malster or corn monger, "a rich priest," who is the solitary capitalist in a community of peasants and artisans. Naturally, he is apt to become their master.[25]

In such circumstances it is not surprising that there should have been a popular outcry against extortion. Inspired by practical grievances, it found an ally, eloquent, if disarmed, in the teaching of the Church. The doctrine as to the ethics of economic conduct, which had been formulated by medieval Popes and interpreted by medieval Schoolmen, was rehearsed by the English divines of the sixteenth century, not merely as the conventional tribute paid by a formal piety to the wisdom of the past, but because the swift changes of the period in commerce and agriculture had, not softened, but accentuated, the problems of conduct for which it had been designed. Nor was it only against the particular case of the covetous money-lender that the preacher and the moralist directed their arrows. The essence of the medieval scheme of economic ethics had been its insistence on equity in bargaining—a contract is fair, St. Thomas had said, when both parties gain from it equally. The prohibition of usury had been the kernel of its doctrines, not because the gains of the money-lender were the only species, but because, in the economic conditions of the age, they were the most conspicuous species, of extortion.

In reality, alike in the Middle Ages and in the sixteenth century, the word usury had not the specialized sense which it carries today. Like the modern profiteer, the usurer was a character so unpopular that most unpopular characters could be called usurers, and by the average practical man almost any form of bargain which he thought oppressive would be classed as usurious. The interpretation placed on the word by those who expounded ecclesiastical theories of usury was equally elastic. Not only the taking of interest for a loan, but the raising of prices by a monopolist, the beating down of prices by a keen bargainer, the rack-renting of land by a landlord, the sub-letting of land by a tenant at a rent higher than he himself paid, the cutting of wages and the paying of wages in truck, the refusal of discount to a tardy debtor, the insistence on unreasonably good security for a loan, the excessive profits of a middleman—all these had been denounced as usury in the very practical thirteenth-century manual of St. Raymond; [26] all these were among the "unlawful chaffer," the "sublety and sleight," which was what the plain man who sat on juries and listened to sermons in parish churches meant by usury three centuries later. If he had been asked why usury was wrong, he would probably have answered with a quotation from Scripture. If he had been asked for a definition of usury, he would have been puzzled, and would have replied in the words of a member of Parliament who spoke on the bill introduced in 1571: "It standeth doubtful what usury is; we have no true definition of it." [27] The truth is, indeed, that any bargain, in which one party obviously gained more advantage than the other, and used his power to the full, was regarded as usurious. The description which best sums up alike popular sentiment and ecclesiastical teaching is contained in the comprehensive indictment applied by his parishioners to an unpopular divine who lent at a penny in the shilling—the cry of all poor men since the world began— Dr. Bennet "is a great taker of advantages." [28]

It was the fact that the theory of usury which the divines of the sixteenth century inherited was not an isolated freak of casuistical ingenuity, but one subordinate element in a comprehensive system of social philosophy, which gave its poignancy to the controversy of which it became the center. The passion which fed on its dusty dialectics was fanned by the conviction that the issue at stake was not merely a legal technicality. It was the fate of the whole scheme of medieval thought, which had attempted to treat economic affairs as part of a hierarchy of values, embracing all interests and activities, of which the apex was religion.

If the Reformation was a revolution, it was a revolution which left almost intact both the lower ranges of ecclesiastical organization and the traditional scheme of social thought. The villager who, resisting the temptations of the alehouse, morris dancing or cards, attended his parish church from 1530 to 1560, must have been bewildered by a succession of changes in the appearance of the building and the form of the services. But there was little to make him conscious of any alteration in the social system of which the church was the center, or in the duties which that system imposed upon himself. After, as before, the Reformation, the parish continued to be a community in which religious and social obligations were inextricably intertwined, and it was as a parishioner, rather than as a subject of the secular authority, that he bore his share of public burdens and performed such public functions as fell to his lot. The officers of whom he saw most in the routine of his daily life were the churchwardens. The place where most public business was transacted, and where news of the doings of the great world came to him, was the parish church. The contributions levied from him were demanded in the name of the parish. Such education as was available for his children was often given by the curate or parish schoolmaster. Such training in coöperation with his fellows as he received sprang from common undertakings maintained by the parish, which owned property, received bequests, let out sheep and cattle, advanced money, made large profits by church ales, and occasionally engaged in trade.[29] Membership of the Church and of the State being co-extensive and equally compulsory, the Government used the ecclesiastical organization of the parish for purposes which, in a later age, when the religious, political and economic aspects of life were disentangled, were to be regarded as secular. The pulpit was the channel through which official information was conveyed to the public and the duty of obedience inculcated. It was to the clergy and the parochial organization that the State turned in coping with pauperism, and down to 1597 collectors for the poor were chosen by the churchwardens in conjunction with the parson.

Where questions of social ethics were concerned, the religious thought of the age was not less conservative than its ecclesiastical organization. Both in their view of religion as embracing all sides of life, and in their theory of the particular social obligations which religion involved, the most representative thinkers of the Church of England had no intention of breaking with traditional doctrines. In the rooted suspicion of economic motives which caused them to damn each fresh manifestation of the spirit of economic enterprise as a new form of the sin of covetousness, as in their insistence that the criteria of economic relations and of the social order were to be sought, not in practical expediency, but in truths of which the Church was the guardian and the exponent, the utterances of men of religion in the reign of Elizabeth, in spite of the revolution which had intervened, had more affinity with the doctrines of the Schoolmen than with those which were to be fashionable after the Restoration.

The oppressions of the tyrannous landlord, who used his economic power to drive an unmerciful bargain, were the subject of constant denunciation down to the Civil War. The exactions of middlemen—"merchants of mischief . . . [who] do make all things dear to the buyers, and yet wonderful vile and of small price to many that must needs

set or sell that which is their own honestly come by"—
were pilloried by Lever.[30]    Nicholas Heming, whose treatise
on *The Lawful Use of Riches* became something like a
standard work, expounded the doctrine of the just price,
and swept impatiently aside the argument which pleaded
freedom of contract as an excuse for covetousness: "Cloake
the same by what title you liste, your synne is excedyng
greate. . . . He which hurteth but one man is in a damnable case; what shall bee thought of thee, whiche bryngest
whole householdes to their graves, or at the leaste art a
meanes of their extreame miserie? Thou maiest finde
shiftes to avoide the danger of men, but assuredly thou
shalte not escape the judgemente of God." [31]   Men eminent
among Anglican divines, such as Sandys and Jewel, took
part in the controversy on the subject of usury.   A bishop
of Salisbury gave his blessing to the book of Wilson; an
archbishop of Canterbury allowed Mosse's sharp *Arraignment* to be dedicated to himself; and a clerical pamphleteer
in the seventeenth century produced a catalogue of six bishops and ten doctors of divinity—not to mention numberless
humbler clergy—who had written in the course of the last
hundred years on different aspects of the sin of extortion
in all its manifold varieties.[32]   The subject was still a favorite of the ecclesiastical orator.   The sixteenth-century
preacher was untrammeled by the convention which in a
more fastidious age was to preclude as an impropriety the
discussion in the pulpit of the problems of the market-
place.   "As it belongeth to the magistrate to punishe," wrote
Heming, "so it is the parte of the preachers to reprove
usurie. . . . First, they should earnestly inveigh against
all unlawfull and wicked contractes. . . . Let them . . .
amend all manifest errours in bargaining by ecclesiasticall
discipline . . . Then, if they cannot reforme all abuses
which they shall finde in bargaines, let them take heede that
they trouble not the Churche overmuche, but commende

the cause unto God . . . Last of all, let them with diligence admonishe the ritche men, that they suffer not themselves to be entangled with the shewe of ritches." [33]

"This," wrote an Anglican divine in reference to the
ecclesiastical condemnation of usury, "hath been the generall judgment of the Church for above this fifteene hundred yeeres, without opposition, in this point. Poor sillie
Church of Christ, that could never finde a lawfull usurie
before this golden age wherein we live." [34]    The first fact
which strikes the modern student of this body of teaching
is its continuity with the past.   In its insistence that buying
and selling, letting and hiring, lending and borrowing, are
to be controlled by a moral law, of which the Church is
the guardian, religious opinion after the Reformation did
not differ from religious opinion before it.   The reformers
themselves were conscious, neither of the emancipation from
the economic follies of the age of medieval darkness ascribed
to them in the eighteenth century, nor of the repudiation
of the traditional economic morality of Christendom, which
some writers have held to have been the result of the revolt
from Rome.   The relation in which they conceived themselves to stand to the social theory of the medieval Church
is shown by the authorities to whom they appealed.   "Therefore I would not," wrote Dr. Thomas Wilson, Master of
Requests and for a short time Secretary of State, "have
men altogether to be enemies to the canon lawe, and to con-
dempne every thinge there written, because the Popes were
aucthours of them, as though no good lawe coulde bee made
by them. . . . Nay, I will saye playnely, that there are
some suche lawes made by the Popes as be righte godly,
saye others what they list." [35]    From the lips of a Tudor
official, such sentiments fell, perhaps, with a certain pi-
quancy.   But, in their appeal to the traditional teaching of
the Church, Wilson's words represented the starting point

from which the discussions of social questions still commonly set out.

The Bible, the Fathers and the Schoolmen, the decretals, church councils, and commentators on the canon law—all these, and not only the first, continued to be quoted as decisive on questions of economic ethics by men to whom the theology and government of the medieval Church were an abomination. What use Wilson made of them, a glance at his book will show. The writer who, after him, produced the most elaborate discussion of usury in the latter part of the century prefaced his work with a list of pre-Reformation authorities running into several pages.[36] The author of a practical memorandum on the amendment of the law with regard to money-lending—a memorandum which appears to have had some effect upon policy—thought it necessary to drag into a paper concerned with the chicanery of financiers and the depreciation of sterling by speculative exchange business, not only Melanchthon, but Aquinas and Hostiensis.[37] Even a moralist who denied all virtue whatever to "the decrees of the Pope" did so only the more strongly to emphasize the prohibition of uncharitable dealing contained in the "statutes of holie Synodes and sayings of godlie Fathers, whiche vehemently forbid usurie."[38] Objective economic science was developing in the hands of the experts who wrote on agriculture, trade, and, above all, on currency and the foreign exchanges. But the divines, if they read such works at all, waved them on one side as the intrusion of Mammon into the fold of Christian morality, and by their obstinate obscurantism helped to prepare an intellectual nemesis, which was to discredit their fervent rhetoric as the voice of a musty superstition. For one who examined present economic realities, ten rearranged thrice-quoted quotations from tomes of past economic casuistry. Sermon was piled upon sermon, and treatise upon treatise. The assumption of all is that the traditional teaching of the Church

as to social ethics is as binding on men's consciences after
the Reformation as it had been before it.

Pamphlets and sermons do not deal either with sins
which no one commits or with sins that every one commits,
and the literary evidence is not to be dismissed merely as
pious rhetoric. The literary evidence does not however,
stand alone. Upon the immense changes made by the Reformation in the political and social position of the Church
it is not necessary to enlarge. It became, in effect, one arm
of the State; excommunication, long discredited by abuse,
was fast losing what little terrors it still retained; a clergy
three-quarters of whom, as a result of the enormous transference of ecclesiastical property, were henceforward presented by lay patrons, were not likely to display any excessive independence. But the canon law was nationalized,
not abolished; the assumption of most churchmen throughout the sixteenth century was that it was to be administered;
and the canon law included the whole body of legislation
as to equity in contracts which had been inherited from the
Middle Ages. True, it was administered no longer by the
clergy acting as the agents of Rome, but by civilians acting
under the authority of the Crown. True, after the prohibition of the study of canon law—after the estimable Dr. Lay-
ton had "set Dunce in Bocardo" at Oxford—it languished
at the universities. True, for the seven years from 1545
to 1552, and again, and on this occasion for good, after
1571, parliamentary legislation expressly sanctioned loans
at interest, provided that it did not exceed a statutory maximum. But the convulsion which changed the source of
canon law did not, as far as these matters are concerned,
alter its scope. Its validity was not the less because it was
now enforced in the name, not of the Pope, but of the King.

As Maitland has pointed out,[39] there was a moment towards the middle of the century when the civil law was
pressing the common law hard. The civil law, as Sir

Thomas Smith assured the yet briefless barrister, offered a promising career, since it was practiced in the ecclesiastical courts.[40]   Though it did not itself forbid usury, it had much to say about it; it was a doctor of the civil law under Elizabeth by whom the most elaborate treatise on the subject was compiled.[41]   By an argument made familiar by a modern controversy on which lay and ecclesiastical opinion have diverged, it is argued that the laxity of the State does not excuse the consciences of men who are the subjects, not only of the State, but of the Church.   "The permission of the Prince," it was urged, "is no absolution from the authority of the Church.   Supposing usury to be unlawfull . . . yet the civil laws permit it, and the Church forbids it.   In this case the Canons are to be preferred. . . . By the laws no man is compelled to be an usurer; and therefore he must pay that reverence and obedience which is otherwise due to them that have the rule over them in the conduct of their souls." [42]

It was this theory which was held by almost all the ecclesiastical writers who dealt with economic ethics in the sixteenth century.   Their view was that, in the words of a pamphleteer, "by the laws of the Church of England . . . usury is simply and generally prohibited." [43]   When the lower House of Convocation petitioned the bishops in 1554 for a restoration of their privileges, they urged, among other matters, that "usurers may be punished by the canon lawes as in tymes past has been used." [44]   In the abortive scheme for the reorganization of the ecclesiastical jurisdiction drawn up by Cranmer and Foxe, usury was included in the list of offenses with which the ecclesiastical courts were to deal, and, for the guidance of judges in what must often have been somewhat knotty cases, a note was added, explaining that it was not to be taken as including the profits derived from objects which yielded increase by the natural process of growth.[45]   Archbishop Grindal's injunctions to

the laity of the Province of York (1571) expressly emphasized the duty of presenting to the Ordinary those who lend
and demand back more than the principal, whatever the
guise under which the transaction may be concealed.[46]
Bishops' articles of visitation down to the Civil War required the presentation of uncharitable persons and usurers,
together with drunkards, ribalds, swearers and sorcerers.[47]
The rules to be observed in excommunicating the impeni-
tent promulgated in 1585, the Canons of the Province of
Canterbury in 1604, and of the Irish Church in 1634, all
included a provision that the usurer should be subjected to
ecclesiastical discipline.[48]

The activity of the ecclesiastical courts had not ceased with
the Reformation, and they continued throughout the last
half of the century to play an important, if increasingly unpopular, part in the machinery of local government.   In
addition to enforcing the elementary social obligation of
charity, by punishing the man who refused to "pay to the
poor men's box," or who was "detected for being an uncharitable person and for not giving to the poor and impotent," [49] they dealt also, at least in theory, with those
who offended against Christian morality by acts of extortion.
The jurisdiction of the Church in these matters was expressly reserved by legislation, and ecclesiastical lawyers,
while lamenting the encroachments of the common law
courts, continued to claim certain economic misdemeanors
as their province.   That, in spite of the rising tide of opposition, the references to questions of this kind in articles of
visitation were not wholly an affair of common form, is
suggested by the protests against the interference of the
clergy in matters of business, and by the occasional cases
which show that commercial transactions continued to be
brought before the ecclesiastical courts.   The typical usurer
was apt, indeed, to outrage not one, but all, of the decencies
of social intercourse.   "Thomas Wilkoxe," complained his

fellow burgesses, "is excommunicated, and disquieteth the parish in the time of divine service. He is a horrible usurer, taking 1*d*. and sometimes 2*d*. for a shilling by the week. He has been cursed by his own father and mother. For the space of two years he hath not received the Holy Communion, but every Sunday, when the priest is ready to go to the Communion, then he departeth the church for the receiving of his weekly usury, and doth not tarry the end of divine service thrice in the year." [50]   Whether the archdeacon corrected a scandal so obviously suitable for ecclesiastical discipline, we do not know.   But in 1578 a case of clerical usury is heard in the court of the archdeacon of Essex.[51]  Twenty-two years later, a usurer is presented with other offenders on the occasion of the visitation of some Yorkshire parishes.[52]   Even in 1619 two instances occur in which money-lenders are cited before the Court of the Commissary of the Bishop of London, on the charge of "lending upon pawnes for an excessive gain commonly reported and cried out of."   One is excommunicated and afterwards absolved; both are admonished to amend their ways.[53]

There is no reason, however, to suppose that such cases were other than highly exceptional; nor is it from the occasional activities of the ever more discredited ecclesiastical jurisdiction that light on the practical application of the ideas of the age as to social ethics is to be sought.   Ecclesiastical discipline is at all times but a misleading clue to the influence of religious opinion, and on the practice of a time when, except for the Court of High Commission, the whole system was in decay, the scanty proceedings of the courts christian throw little light.   To judge the degree to which the doctrines expounded by divines were accepted or repudiated by the common sense of the laity, one must turn to the records which show how questions of business ethics were handled by individuals, by municipal bodies and by the Government.

The opinion of the practical man on questions of economic conduct was in the sixteenth century in a condition of even more than its customary confusion. A century before, he had practised extortion and been told that it was wrong; for it was contrary to the law of God. A century later, he was to practise it and be told that he was right; for it was in accordance with the law of nature. In this matter, as in others of even greater moment, the two generations which followed the Reformation were unblessed by these ample certitudes. They walked in an obscurity where the glittering armor of theologians

made
A little glooming light, most like a shade.

In practice, since new class interests and novel ideas had arisen, but had not yet wholly submerged those which preceded them, every shade of opinion, from that of the pious burgess, who protested indignantly against being saddled with a vicar who took a penny in the shilling, to the latitudinarianism of the cosmopolitan financier, to whom the confusion of business with morals was a vulgar delusion, was represented in the economic ethics of Elizabethan England. As far as the smaller property-owners were concerned, the sentiment of laymen differed, on the whole, less widely from the doctrines expounded by divines, than it did from the individualism which was beginning to carry all before it among the leaders of the world of business. Against the rising financial interests of the day were arrayed the stolid conservatism of the peasantry and the humbler *bourgeoisie,* whose conception of social expediency was the defence of customary relations against innovation, and who regarded the growth of this new power with something of the same jealous hostility as they opposed to the economic radicalism of the enclosing landlord. At bottom, it was an instinctive movement of self-protection. Free play for the capitalist seemed to menace the independence of the small producer, who tilled the nation's fields and wove its cloth. The path down which the financier beguiles his victims may seem at first to be strewn with roses; but at the end of it lies—incredible nightmare—a *régime* of universal capitalism, in which peasant and small master will have been merged in a property-less proletariat, and "the riches of the city of London, and in effect of all this realm, shall be at that time in the hands of a few men having unmerciful hearts." [54]

Against the landlord who enclosed commons, converted arable to pasture, and rack-rented his tenants, local resentment, unless supported by the Government, was powerless. Against the engrosser, however, it mobilized the traditional machinery of maximum prices and market regulations, and dealt with the usurer as best it could, by presenting him before the justices in Quarter Sessions, by advancing money from the municipal exchequer to assist his victims, and even, on occasion, by establishing a public pawnshop, with a monopoly of the right to make loans, as a protection to the inhabitants against extreme "usurers and extortioners." The commonest charity of the age, which was the establishment of a fund to make advances without interest to tradesmen, was inspired by similar motives. Its aim was to enable the young artisan or shopkeeper, the favorite victim of the money-lender, to acquire the indispensable "stock," without which he could not set up in business. [55]

The issues which confronted the Government were naturally more complicated, and its attitude was more ambiguous. The pressure of commercial interests growing in wealth and influence, its own clamorous financial necessities, the mere logic of economic development, made it out of the question for it to contemplate, even if it had been disposed to do so, the rigorous economic discipline desired by the divines. Tradition, a natural conservatism, the apprehension of public disorder caused by enclosures or by distress among the industrial population, a belief in its own mission as the guardian of "good order" in trade, not unmingled with a hope that the control of economic affairs might be made to yield agreeable financial pickings, gave it a natural bias to a policy which aimed at drawing all the threads of economic life into the hands of a paternal monarchy.

In the form which the system assumed under Elizabeth, considerations of public policy, which appealed to the State, were hardly distinguishable from considerations of social morality, which appealed to the Church. As a result of the Reformation the relations previously existing between the Church and the State had been almost exactly reversed. In the Middle Ages the former had been, at least in theory, the ultimate authority on questions of public and private morality, while the latter was the police-officer which enforced its decrees. In the sixteenth century, the Church became the ecclesiastical department of the State, and religion was used to lend a moral sanction to secular social policy. But the religious revolution had not destroyed the conception of a single society, of which Church and State were different aspects; and, when the canon law became "the King's ecclesiastical law of England," the jurisdiction of both inevitably tended to merge. Absorbing the ecclesiastical authority into itself, the Crown had its own reasons of political expediency for endeavoring to maintain traditional standards of social conduct, as an antidote for what Cecil called "the license grown by liberty of the Gospel." Ecclesiastics, in their turn, were public officers—under Elizabeth the bishop was normally also a justice of the peace—and relied on secular machinery to enforce, not only religious conformity, but Christian morality, because both were elements in a society in which secular and spiritual interests had not yet been completely disentangled from each other. "We mean by the Commonwealth," wrote Hooker, "that society with relation unto all public affairs thereof, only the matter of true religion accepted; by the Church, the same society, with only reference unto the matter of true religion, without any other affairs besides." [56]

In economic and social, as in ecclesiastical, matters, the opening years of Elizabeth were a period of conservative reconstruction. The psychology of a nation which lives predominantly by the land is in sharp contrast with that of a commercial society. In the latter, when all goes well, continuous expansion is taken for granted as the rule of life, new horizons are constantly opening, and the catchword of politics is the encouragement of enterprise. In the former, the number of niches into which each successive generation must be fitted is strictly limited; movement means disturbance, for, as one man rises, another is thrust down; and the object of statesmen is, not to foster individual initiative, but to prevent social dislocation. It was in this mood that Tudor Privy Councils approached questions of social policy and industrial organization. Except when they were diverted by financial interests, or lured into ambitious, and usually unsuccessful, projects for promoting economic development, their ideal was, not progress, but stability. Their enemies were disorder, and the restless appetites which, since they led to the encroachment of class on class, were thought to provoke it. Distrusting economic individualism for reasons of state as heartily as did churchmen for reasons of religion, their aim was to crystallize existing class relationships by submitting them to the pressure, at once restrictive and protective, of a paternal Government, vigilant to detect all movements which menaced the established order, and alert to suppress them.

Take but degree away, untune that string,
And, hark, what discord follows! . . .
Force should be right; or rather, right and wrong
(Between whose endless jar justice resides)
Should lose their names, and so should justice too.
Then every thing includes itself in power,
Power into will, will into appetite;
And appetite, an universal wolf,
So doubly seconded with will and power,
And, last, eat up himself.

In spite of the swift expansion of commerce in the latter part of the century, the words of Ulysses continued for long to express the official attitude.

The practical application of such conceptions was an elaborate system of what might be called, to use a modern analogy, "controls." Wages, the movement of labor, the entry into a trade, dealings in grain and in wool, methods of cultivation, methods of manufacture, foreign exchange business, rates of interest—all are controlled, partly by Statute, but still more by the administrative activity of the Council. In theory, nothing is too small or too great to escape the eyes of an omniscient State. Does a landowner take advantage of the ignorance of peasants and the uncertainty of the law to enclose commons or evict copyholders? The Council, while protesting that it does not intend to hinder him from asserting his rights at common law, will intervene to stop gross cases of oppression, to prevent poor men from being made the victims of legal chicanery and intimidation, to settle disputes by common sense and moral pressure, to remind the aggressor that he is bound "rather to consider what is agreeable . . . to the use of this State and for the good of the comon wealthe, than to seeke the uttermost advantage that a landlord for his particular profit maie take amonge his tenaunts." [57] Have prices been raised by a bad harvest? The Council will issue a solemn denun-

ciation of the covetousness of speculators, "in conditions
more like to wolves or cormorants than to natural men," [58]
who take advantage of the dearth to exploit public necessities; will instruct the Commissioners of Grain and Victuals
to suspend exports; and will order justices to inspect barns,
ration supplies, and compel farmers to sell surplus stocks
at a fixed price. Does the collapse of the continental market threaten distress in the textile districts? The Council
will put pressure on clothiers to find work for the opera-
tives, "this being the rule by which the wool-grower, the
clothier and merchant must Le governed, that whosoever
had a part of the gaine in profitable times . . . must now,
in the decay of trade . . . beare a part of the publicke
losses, as may best conduce to the good of the publicke and
the maintenance of the generall trade." [59] Has the value of
sterling fallen on the Antwerp market? The Council will
consider pegging the exchanges, and will even attempt to
nationalize foreign exchange business by prohibiting private transactions altogether.[60] Are local authorities negligent in the administration of the Poor Law? The Council,
which insists on regular reports as to the punishment of
vagrants, the relief of the impotent, and the steps taken to
provide materials on which to employ the able-bodied, inun-
dates them with exhortations to mend their ways and with
threats of severer proceedings if they fail. Are tradesmen
in difficulties? The Council, which keeps sufficiently in
touch with business conditions to know when the difficulties of borrowers threaten a crisis, endeavors to exercise a
moderating influence by making an example of persons
guilty of flagrant extortion, or by inducing the parties to
accept a compromise. A mortgagee accused of "hard and
unchristianly dealing" is ordered to restore the land which
he has seized, or to appear before the Council. A creditor
who has been similarly "hard and unconscionable" is committed to the Fleet. The justices of Norfolk are instructed

to put pressure on a money-lender who has taken "very unjust and immoderate advantage by way of usury." The bishop of Exeter is urged to induce a usurer in his diocese to show "a more Christian and charitable consideration of these his neighbors." A nobleman has released two offenders imprisoned by the High Commission for the Province of York for having "taken usury contrary to the laws of God and of the realm," and is ordered at once to recommit them. No Government can face with equanimity a state of things in which large numbers of respectable tradesmen may be plunged into bankruptcy. In times of unusual depression, the Council's intervention to prevent creditors from pressing their claims to the hilt was so frequent as to create the impression of something like an informal moratorium.[61]

The Governments of the Tudors and, still more, of the first two Stuarts, were masters of the art of disguising commonplace, and sometimes sordid, motives beneath a glittering façade of imposing principles. In spite of its lofty declarations of a disinterested solicitude for the public welfare, the social policy of the monarchy not only was as slipshod in execution as it was grandiose in design, but was not seldom perverted into measures disastrous to its ostensible ends, both by the sinister pressure of sectional interests, and by the insistent necessities of an empty exchequer. Its fundamental conception, however—the philosophy of the thinkers and of the few statesmen who rose above immediate exigencies to consider the significance of the system in its totality—had a natural affinity with the doctrines which commended themselves to men of religion. It was of an ordered and graded society, in which each class performed its allotted function, and was secured such a livelihood, and no more than such a livelihood, as was proportioned to its status. "God and the Kinge," wrote one who had labored much, amid grave personal dangers, for the

welfare of his fellows, "hathe not sent us the poore lyvinge
we have, but to doe services therfore amonge our neighbours abroade." [62]   The divines who fulminated against the
uncharitable covetousness of the extortionate middleman,
the grasping money-lender, or the tyrannous landlord, saw
in the measures by which the Government endeavored to
suppress the greed of individuals or the collision of classes
a much needed cement of social solidarity, and appealed to
Cæsar to redouble his penalties upon an economic license
which was hateful to God.   The statesmen concerned to
prevent agitation saw in religion the preservative of order,
and the antidote for the cupidity or ambition which threatened to destroy it, and reënforced the threat of temporal
penalties with arguments that would not have been out of
place in the pulpit.   To both alike religion is concerned with
something more than personal salvation.   It is the sanction
of social duties and the spiritual manifestation of the corporate life of a complex, yet united, society.   To both the
State is something more than an institution created by material necessities or political convenience.   It is the temporal expression of spiritual obligations.   It is a link between the individual soul and that supernatural society of
which all Christian men are held to be members.   It rests
not merely on practical convenience, but on the will of God.

Of that philosophy, the classical expression, at once the
most catholic, the most reasonable and the most sublime,
is the work of Hooker.   What it meant to one cast in a
narrower mould, pedantic, irritable and intolerant, yet not
without the streak of harsh nobility which belongs to all
who love an idea, however unwisely, more than their own
ease, is revealed in the sermons and the activity of Laud.
Laud's intellectual limitations and practical blunders need
no emphasis.   If his vices made him intolerable to the most
powerful forces of his own age, his virtues were not of a
kind to commend him to those of its successor, and history

has been hardly more merciful to him than were his political opponents. But an intense conviction of the fundamental solidarity of all the manifold elements in a great community, a grand sense of the dignity of public duties, a passionate hatred for the self-seeking pettiness of personal cupidities and sectional interests—these qualities are not among the weaknesses against which the human nature of ordinary men requires to be most upon its guard, and these qualities Laud possessed, not only in abundance, but to excess. His worship of unity was an idolatry, his detestation of faction a superstition. Church and State are one Jerusalem: "Both Commonwealth and Church are collective bodies, made up of many into one; and both so near allied that the one, the Church, can never subsist but in the other, the Commonwealth; nay, so near, that the same men, which in a temporal respect make the Commonwealth, do in a spiritual make the Church." [63] Private and public interests are inextricably interwoven. The sanction of unity is religion. The foundation of unity is justice: "God will not bless the State, if kings and magistrates do not execute judgment, if the widow and the fatherless have cause to cry out against the 'thrones of justice.' " [64]

To a temper so permeated with the conception that society is an organism compact of diverse parts, and that the grand end of government is to maintain their coöperation, every social movement or personal motive which sets group against group, or individual against individual, appears, not the irrepressible energy of life, but the mutterings of chaos. The first demon to be exorcised is party, for Governments must "entertain no private business," and "parties are ever private ends." [65] The second is the self-interest which leads the individual to struggle for riches and advancement. "There is no private end, but in something or other it will be led to run cross the public; and, if gain come in, though it be by 'making shrines for Diana,' it is no matter with them though Ephesus be in an uproar for it." [66]    For Laud, the political virtues, by which he understands subordination, obedience, a willingness to sacrifice personal interests for the good of the community, are as much part of the Christian's religion as are the duties of private life; and, unlike some of those who sigh for social unity today, he is as ready to chastise the rich and powerful, who thwart the attainment of that ideal, as he is to preach it to the humble.    To talk of holiness and to practice injustice is mere hypocrisy. Man is born a member of a society and is dedicated by religion to the service of his fellows.    To repudiate the obligation is to be guilty of a kind of political atheism.

"If any man be so addicted to his private, that he neglect the common, state, he is void of the sense of piety and wisheth peace and happiness to himself in vain.    For whoever he be, he must live in the body of the Commonwealth, and in the body of the Church." [67]    To one holding such a creed economic individualism was hardly less abhorrent than religious nonconformity, and its repression was a not less obvious duty; for both seemed incompatible with the stability of a society in which Commonwealth and Church were one.    It is natural, therefore, that Laud's utterances and activities in the matter of social policy should have shown a strong bias in favor of the control of economic relations by an authoritarian State, which reached its climax in the eleven years of personal government.    It was a moment when, partly in continuance of the traditional policy of protecting peasants and maintaining the supply of grain, partly for less reputable reasons of finance, the Government was more than usually active in harrying the depopulating landlord.    The Council gave sympathetic consideration to petitions from peasants begging for protection or redress, and in 1630 directions were issued to the justices of five midland counties to remove all enclosures made in the last five years, on the ground that they resulted in depopulation and were particularly harmful in times of dearth. In 1632, 1635, and 1636, three Commissions were appointed and special instructions against enclosure were issued to the Justices of Assize. In parts of the country, at any rate, land which had been laid down to grass was plowed up in obedience to the Government's orders. In the four years from 1635 to 1638 a list of some 600 offenders was returned to the Council, and about £50,000 was imposed upon them in fines.[68] With this policy Laud was whole-heartedly in sympathy. A letter in his private correspondence, in which he expresses his detestation of enclosure, reveals the temper which evoked Clarendon's gentle complaint that the archbishop made himself unpopular by his inclination "a little too much to countenance the Commission for Depopulation." [69] Laud was himself an active member of the Commission, and dismissed with impatient contempt the squirearchy's appeal to the common law. In the day of his ruin he was reminded by his enemies of the needlessly sharp censures with which he barbed the fine imposed upon an enclosing landlord.[70]

The prevention of enclosure and depopulation was merely one element in a general policy, by which a benevolent Government, unhampered by what Laud had called "that noise" of parliamentary debate, was to endeavor by even-handed pressure to enforce social obligations on great and small, and to prevent the public interest being sacrificed to an unconscionable appetite for private gain. The preoccupation of the Council with the problem of securing adequate food supplies and reasonable prices, with poor relief, and, to a lesser degree, with questions of wages, has been described by Miss Leonard, and its attempts to protect craftsmen against exploitation at the hands of merchants by Professor Unwin.[71] In 1630-1 it issued in an amended form the Elizabethan Book of Orders, instructing justices as to their duty to see that markets were served and prices controlled, appointed a special committee of the Privy Council as Commissioners of the Poor and later a separate Commission, and issued a Book of Orders for the better administration of the Poor Law. In 1629, 1631, and again in 1637, it took steps to secure that the wages of textile workers in East Anglia were raised, and punished with imprisonment in the Fleet an employer notorious for paying in truck. As President of the Council of the North, Wentworth protected the commoners whose vested interests were threatened by the drainage of Hatfield Chase, and endeavored to insist on the stricter administration of the code regulating the woollen industry.[72]

Such action, even if inspired largely by the obvious interest of the Government, which had enemies enough on its hands already, in preventing popular discontent, was of a kind to appeal to one with Laud's indifference to the opinion of the wealthier classes, and with Laud's belief in the divine mission of the House of David to teach an obedient people "to lay down the private for the public sake." It is not surprising, therefore, when the Star Chamber fines an engrosser of corn, to find him improving the occasion with the remark that the defendant has been "guilty of a most foule offence, which the Prophet hath [called] in a very energeticall phrase grynding the faces of the poore," and that the dearth has been caused, not by God, but by "cruell men"; [73] or taking part in the proceedings of the Privy Council at a time when it is pressing justices, apparently not without success, to compel the East Anglian clothiers to raise the wages of spinners and weavers; or serving on the Lincolnshire sub-committee of the Commission on the Relief of the Poor, which was appointed in January 1631.[74]

"A bishop," observed Laud, in answer to the attack of Lord Saye and Sele, "may preach the Gospel more publicly and to far greater edification in a court of judicature, or at a Council-table, where great men are met together to draw things to an issue, than many preachers in their several charges can." [75] The Church, which had abandoned the pretension itself to control society, found some compensation in the reflection that its doctrines were not wholly without influence in impressing the principles which were applied by the State. The history of the rise of individual liberty —to use a question-begging phrase—in economic affairs follows somewhat the same course as does its growth in the more important sphere of religion, and is not unconnected with it. The conception of religion as a thing private and individual does not emerge until after a century in which religious freedom normally means the freedom of the State to prescribe religion, not the freedom of the individual to worship God as he pleases. The assertion of economic liberty as a natural right comes at the close of a period in which, while a religious phraseology was retained and a religious interpretation of social institutions was often sincerely held, the supernatural sanction had been increasingly merged in doctrines based on reasons of state and public expediency. "Jerusalem . . . stands not for the City and the State only . . . nor for the Temple and the Church only, but jointly for both." [76] In identifying the maintenance of public morality with the spasmodic activities of an incompetent Government, the Church had built its house upon the sand. It did not require prophetic gifts to foresee that the fall of the City would be followed by the destruction of the Temple.

### III. THE GROWTH OF INDIVIDUALISM

Though the assertion of the traditional economic ethics continued to be made by one school of churchmen down to the meeting of the Long Parliament, it was increasingly the voice of the past appealing to an alien generation. The expression of a theory of society which had made religion supreme over all secular affairs, it had outlived the synthesis in which it had been an element, and survived, an archaic fragment, into an age to whose increasing individualism the idea of corporate morality was as objectionable as that of ecclesiastical discipline by bishops and archdeacons was becoming to its religion. The collision between the prevalent practice, and what still purported to be the teaching of the Church, is almost the commonest theme of the economic literature of the period from 1550 to 1640; of much of it, indeed, it is the occasion. Whatever the Church might say, men had asked interest for loans, and charged what prices the market would stand, at the very zenith of the Age of Faith. But then, except in the great commercial centers and in the high finance of the Papacy and of secular Governments, their transactions had been petty and individual, an occasional shift to meet an emergency or seize an opportunity. The new thing in the England of the sixteenth century was that devices that had formerly been occasional were now woven into the very texture of the industrial and commercial civilization which was developing in the later years of Elizabeth, and whose subsequent enormous expansion was to give English society its characteristic quality and tone. Fifty years later, Harrington, in a famous passage, described how the ruin of the feudal nobility by the Tudors, by democratizing the ownership of land, had prepared the way for the *bourgeois* republic.[77] His hint of the economic changes which preceded the Civil War might be given a wider application. The age of Elizabeth saw a steady growth of capitalism in textiles and mining, a great increase of foreign trade and an outburst of joint-stock enterprise in connection with it, the beginnings of something like deposit banking in the hands of the scriveners, and the growth, aided by the fall of Antwerp and the Government's own financial necessities, of a money-market with an almost modern technique—speculation, futures and arbitrage transactions—in London. The future lay with the classes who sprang to wealth and influence with the expansion of commerce in the later years of the century, and whose religious and political aspirations were, two generations later, to overthrow the monarchy.

An organized money-market has many advantages. But it is not a school of social ethics or of political responsibility. Finance, being essentially impersonal, a matter of opportunities, security and risks, acted among other causes as a solvent of the sentiment, fostered both by the teaching of the Church and the decencies of social intercourse among neighbors, which regarded keen bargaining as "sharp practice." In the half-century which followed the Reformation, thanks to the collapse of sterling on the international market, as a result of a depreciated currency, war, and a foreign debt contracted on ruinous terms, the state of the foreign exchanges was the obsession of publicists and politicians. Problems of currency and credit lend themselves more readily than most economic questions to discussion in terms of mechanical causation. It was in the long debate provoked by the rise in prices and the condition of the exchanges, that the psychological assumptions, which were afterwards to be treated by economists as of self-evident and universal validity, were first hammered out.

"We see," wrote Malynes, "how one thing driveth or enforceth another, like as in a clock where there are many wheels, the first wheel being stirred driveth the next and that the third and so forth, till the last that moveth the instrument that striketh the clock; or like as in a press going in a strait, where the foremost is driven by him that is next to him, and the next by him that followeth him." [78] The spirit of modern business could hardly be more aptly described. Conservative writers denounced it as fostering a soulless individualism, but, needless to say, their denunciations were as futile as they were justified. It might be possible to put fear into the heart of the village dealer who bought cheap and sold dear, or of the pawnbroker who took a hundred quarters of wheat when he had lent ninety, with the warning that "the devices of men cannot be concealed from Almighty God." To a great clothier, or to a capitalist like Pallavicino, Spinola, or Thomas Gresham, who managed the Government business in Antwerp, such sentiments were foolishness, and usurious interest appeared, not bad morals, but bad business. Moving, as they did, in a world where loans were made, not to meet the temporary difficulty of an unfortunate neighbor, but as a profitable investment on the part of not too scrupulous business men, who looked after themselves and expected others to do the same, they had scanty sympathy with doctrines which reflected the spirit of mutual aid not unnatural in the small circle of neighbors who formed the ordinary village or borough in rural England.

It was a natural result of their experience that, without the formal enunciation of any theory of economic individualism, they should throw their weight against the traditional restrictions, resent the attempts made by preachers and popular movements to apply doctrines of charity and "good conscience" to the impersonal mechanism of large-scale transactions, and seek to bring public policy more into accordance with their economic practice. The opposition to the Statutes against depopulation offered by the self-interest of the gentry was being supported in the latter years of Elizabeth by free-trade arguments in the House of Commons, and the last Act, which was passed in 1597, expressly allowed land to be laid down to pasture for the purpose of giving it a rest.[79] From at any rate the middle of the century, the fixing of prices by municipal authorities and by the Government was regarded with skepticism by the more advanced economic theorists, and towards the end of the century it produced complaints that, since it weakened the farmer's incentive to grow corn, its results were the precise opposite of those intended.[80]   As markets widened, the control of the middleman who dealt in wool and grain, though strictly enforced in theory, showed unmistakable signs of breaking down in practice.   Gresham attacked the prohibition of usury, and normally stipulated that financiers who subscribed on his inducement to public loans should be indemnified against legal proceedings.[81]   Nor could he well have done otherwise, for the sentiment of the City was that of the merchant in Wilson's Dialogue: "What man is so madde to deliver his moneye out of his owne possession for naughte? or whoe is he that will not make of his owne the best he can?" [82]   With such a wind of doctrine in their sails men were not far from the days of complete freedom of contract.

Most significant of all, economic interests were already appealing to the political theory which, when finally systematized by Locke, was to prove that the State which interferes with property and business destroys its own title to exist.   "All free subjects," declared a Committee of the House of Commons in 1604, "are born inheritable, as to their land, so also to the free exercise of their industry, in those trades whereto they apply themselves and whereby they are to live.   Merchandise being the chief and richest of all other, and of greater extent and importance than all the rest, it is against the natural right and liberty of the subjects of England to restrain it into the hands of some few." [83]   The process by which natural justice, imperfectly embodied in positive law, was replaced as the source of authority by positive law which might or might not be the expression of natural justice, had its analogy in the rejection by social theory of the whole conception of an objective standard of economic equity.   The law of nature had been invoked by medieval writers as a moral restraint upon economic self-interest.   By the seventeenth century, a significant revolution had taken place. "Nature" had come to connote, not divine ordinance, but human appetites, and natural rights were invoked by the individualism of the age as a reason why self-interest should be given free play.

The effect of these practical exigencies and intellectual changes was seen in a reversal of policy on the part of the State. In 1571 the Act of 1552, which had prohibited all interest as "a vyce moste odyous and detestable, as in dyvers places of the hollie Scripture it is evydent to be seen," had been repealed, after a debate in the House which revealed the revolt of the plain man against the theorists who had triumphed twenty years before, and his determination that the law should not impose on business a utopian morality.[84] The exaction of interest ceased to be a criminal offence, provided that the rate did not exceed ten per cent., though it still remained open to a debtor, in the improbable event of his thinking it expedient to jeopardize his chance of future advances, to take civil proceedings to recover any payment made in excess of the principal. This qualified condonation of usury on the part of the State naturally reacted upon religious opinion. The Crown was supreme ruler of the Church of Christ, and it was not easy for a loyal Church to be more fastidious than its head. Moderate interest, if without legal protection, was at any rate not unlawful, and it is difficult to damn with conviction vices of which the degrees have been adjusted on a sliding scale by an Act of Parliament. Objective economic science was beginning its disillusioning career, in the form of discussions on the rise in prices, the mechanism of the money-market, and the balance of trade, by publicists concerned, not to point a moral, but to analyze forces so productive of profit to those interested in their operation. Since Calvin's indulgence to interest, critics of the traditional doctrine could argue that religion itself spoke with an uncertain voice.

Such developments inevitably affected the tone in which

the discussion of economic ethics was carried on by the divines, and even before the end of the sixteenth century, though they did not dream of abandoning the denunciation of unconscionable bargains, they were surrounding it with qualifications.   The *Decades* of Bullinger, of which three English translations were made in the ten years following his death, and which Convocation in 1586 required to be obtained and studied by all the inferior clergy, indicated a *via media*.   As uncompromising as any medieval writer in his hatred of the sin of covetousness, he denounces with all the old fervor oppressive contracts which grind the poor.   But he is less intolerant of economic motives than most of his predecessors, and concedes, with Calvin, that, before interest is condemned as usury, it is necessary to consider both the terms of the loan and the position of borrower and lender.

The stricter school of religious opinion continued to cling to the traditional theory down to the Civil War.   Conservative divines took advantage of the section in the Act of 1571 declaring that "all usurie being forbydden by the lawe of God is synne and detestable," to argue that the Statute had in reality altered nothing, and that the State left it to the Church to prevent bargains which, for reasons of practical expediency, it did not think fit to prohibit, but which it did not encourage and declined to enforce.   It is in obedience to such doctrines that a scrupulous parson refuses a cure until he is assured that the money which will be paid to him comes from the rent of land, not from interest on capital.[85]   But, even so, there are difficulties.   The parson of Kingham bequeaths a cow to the poor of Burford, which is "set to hire for a year or two for four shillings a year," the money being used for their assistance.   But the arrangement has its inconveniences.   Cows are mortal, and this communal cow is "very like to have perished through casualty and ill-keeping." [86]   Will not the poor be surer of their money if the cow is disposed of for cash down?  So it is sold to the man who previously hired it, and the interest spent on the poor instead.  Is this usury?  Is it usury to invest money in business in order to provide an income for those, like widows and orphans, who cannot trade with it themselves?  If it is lawful to buy a rent-charge or to share in trading profits, what is the particular criminality of charging a price for a loan?  Why should a creditor, who may himself be poor, make a loan *gratis,* in order to put money into the pocket of a wealthy capitalist, who uses the advance to corner the wool crop or to speculate on the exchanges?

To such questions liberal theologians answered that the crucial point was not the letter of the law which forbad the breeding of barren metal, but the observance of Christian charity in economic, as in other, transactions.  Their opponents appealed to the text of Scripture and the law of the Church, argued that usury differed, not merely in degree, but in kind, from payments which, like rent and profits, were morally unobjectionable provided that they were not extortionate in amount, and insisted that usury was to be interpreted as "whatever is taken for a loan above the principal."  The literature of the subject was voluminous.  But it was obsolete almost before it was produced.  For, whether theologians and moralists condemned all interest, or only some interest, as contrary to Christian ethics, the assumption implied in their very disagreement had been that economic relations belonged to a province of which, in the last resort, the Church was master.  That economic transactions were one department of ethical conduct, and to be judged, like other parts of it, by spiritual criteria; that, whatever concessions the State might see fit to make to human frailty, a certain standard of economic morality was involved in membership of the Christian Church; that it was the function of ecclesiastical authorities, whoever they might be, to take the action needed to bring home to men their social obligations—such doctrines were still common ground to all sections of religious thought. It was precisely this whole conception of a social theory based ultimately on religion which was being discredited. While rival authorities were discussing the correct interpretation of economic ethics, the flank of both was turned by the growth of a powerful body of lay opinion, which argued that economics were one thing and ethics another.

Usury, a summary name for all kinds of extortion, was the issue in which the whole controversy over "good conscience" in bargaining came to a head, and such questions were only one illustration of the immense problems with which the rise of a commercial civilization confronted a Church whose social ethics still professed to be those of the Bible, the Fathers and the Schoolmen. A score of books, garnished with citations from Scripture and from the canonists, were written to answer them. Many of them are learned; some are almost readable. But it may be doubted whether, even in their own day, they satisfied any one but their authors. The truth is that, in spite of the sincerity with which it was held that the transactions of business must somehow be amenable to the moral law, the code of practical ethics, in which that claim was expressed, had been forged to meet the conditions of a very different environment from that of commercial England in the seventeenth century.

The most crucial and the most difficult of all political questions is that which turns on the difference between public and private morality. The problem which it presents in the relations between States is a commonplace. But, since its essence is the difficulty of applying the same moral standard to decisions which affect large masses of men as to those in which only individuals are involved, it emerges in a hardly less acute form in the sphere of economic life, as soon as its connections ramify widely, and the unit is no longer the solitary producer, but a group. To argue, in the manner of Machiavelli, that there is one rule for business and another for private life, is to open a door to an orgy of unscrupulousness before which the mind recoils. To argue that there is no difference at all is to lay down a principle which few men who have faced the difficulty in practice will be prepared to endorse as of invariable application, and incidentally to expose the idea of morality itself to discredit by subjecting it to an almost intolerable strain. The practical result of sentimentality is too often a violent reaction towards the baser kinds of *Realpolitik*.

With the expansion of finance and international trade in the sixteenth century, it was this problem which faced the Church. Granted that I should love my neighbor as myself, the questions which, under modern conditions of large-scale organization, remain for solution are, Who precisely *is* my neighbor? and, How exactly am I to make my love for him effective in practice? To these questions the conventional religious teaching supplied no answer, for it had not even realized that they could be put. It had tried to moralize economic relations by treating every transaction as a case of personal conduct, involving personal responsibility. In an age of impersonal finance, world-markets and a capitalist organization of industry, its traditional social doctrines had no specific to offer, and were merely repeated, when, in order to be effective, they should have been thought out again from the beginning and formulated in new and living terms. It had endeavored to protect the peasant and the craftsman against the oppression of the money-lender and the monopolist. Faced with the problems of a wage-earning proletariat, it could do no more than repeat, with meaningless iteration, its traditional lore as to the duties of master to servant and servant to master. It had insisted that all men were brethren. But it did not occur to it to point out that, as a result of the new economic imperialism which was beginning to develop in the seventeenth century, the brethren of the English merchant were the Africans whom he kidnaped for slavery in America, or the American Indians whom he stripped of their lands, or the Indian craftsmen from whom he bought muslins and silks at starvation prices.    Religion had not yet learned to console itself for the practical difficulty of applying its moral principles by clasping the comfortable formula that for the transactions of economic life no moral principles exist.    But, for the problems involved in the association of men for economic purposes on the grand scale which was to be increasingly the rule in the future, the social doctrines advanced from the pulpit offered, in their traditional form, little guidance.    Their practical ineffectiveness prepared the way for their theoretical abandonment.

They were abandoned because, on the whole, they deserved to be abandoned.    The social teaching of the Church had ceased to count, because the Church itself had ceased to think.    Energy in economic action, realist intelligence in economic thought—these qualities were to be the note of the seventeenth century, when once the confusion of the Civil War had died down.    When mankind is faced with the choice between exhilarating activities and piety imprisoned in a shriveled mass of desiccated formulæ, it will choose the former, though the energy be brutal and the intelligence narrow.    In the age of Bacon and Descartes, bursting with clamorous interests and eager ideas, fruitful, above all, in the germs of economic speculation, from which was to grow the new science of Political Arithmetic, the social theory of the Church of England turned its face from the practical world, to pore over doctrines which, had their original authors been as impervious to realities as their later exponents, would never have been formulated.    Naturally it was shouldered aside.  It was neglected because it had become negligible.

The defect was fundamental.  It made itself felt in countries where there was no Reformation, no Puritan movement, no common law jealous of its rights and eager to prune ecclesiastical pretensions.  But in England there were all three, and, from the beginning of the last quarter of the sixteenth century, ecclesiastical authorities who attempted to enforce traditional morality had to reckon with a temper which denied their right to exercise any jurisdiction at all, above all, any jurisdiction interfering with economic matters.  It was not merely that there was the familiar objection of the plain man that parsons know nothing of business—that "it is not in simple divines to show what contract is lawful and what is not." [87]  More important, there was the opposition of the common lawyers to part, at least, of the machinery of ecclesiastical discipline.  Bancroft in 1605 complained to the Privy Council that the judges were endeavoring to confine the jurisdiction of the ecclesiastical courts to testamentary and matrimonial cases, and alleged that, of more than five hundred prohibitions issued to stop proceedings in the Court of Arches since the accession of Elizabeth, not more than one in twenty could be sustained.[88]  "As things are," wrote two years later the author of a treatise on the civil and ecclesiastical law, "neither jurisdiction knowes their owne bounds, but one snatcheth from the other, in maner as in a batable ground lying betweene two kingdomes." [89]  The jurisdiction of the Court of High Commission suffered in the same way.  In the last resort appeals from the ecclesiastical courts went either to it or to the Court of Delegates.  From the latter part of the sixteenth century down to the removal of Coke from the Bench in 1616, the judges were from time to time staying proceedings before the Court of High Commission by prohibitions, or discharging offenders imprisoned by it.  In 1577, for example, they released on a writ of *Habeas Corpus* a prisoner committed by the High Commission on a charge of usury.[90]

Most fundamental of all, there was the growth of a theory of the Church, which denied the very principle of a discipline exercised by bishops and archdeacons. The acquiescence of the laity in the moral jurisdiction of the clergy had been accorded with less and less readiness for two centuries before the Reformation. With the growth under Elizabeth of a vigorous Puritan movement, which had its stronghold among the trading and commercial classes, that jurisdiction became to a considerable proportion of the population little less than abhorrent. Their dislike of it was based, of course, on weightier grounds than its occasional interference in matters of business. But their attitude had as an inevitable result that, with the disparagement of the whole principle of the traditional ecclesiastical discipline, that particular use of it was also discredited. It was not that Puritanism implied a greater laxity in social relations. On the contrary, in its earlier phases it stood, at least in theory, for a stricter discipline of the life of the individual, alike in his business and in his pleasures. But it repudiated as anti-Christian the organs through which such discipline had in fact been exercised. When the Usury Bill of 1571 was being discussed in the House of Commons, reference to the canon law was met by the protest that the rules of the canon law on the matter were abolished, and that "they should be no more remembered than they are followed." [91] Feeling against the system rose steadily during the next two generations; excommunications, when courts ventured to resort to them, were freely disregarded; [92] and by the thirties of the seventeenth century, under the influence of Laud's *régime*, the murmur was threatening to become a hurricane. Then came the Long Parliament, the fierce denunciations in both Houses of the interference of the clergy in civil affairs, and the legislation abolishing the Court of High Commission, depriving the ordinary ecclesiastical courts of penal jurisdiction, and finally, with the abolition of episcopacy, sweeping them away altogether.

"Not many good days," wrote Penn, "since ministers meddled so much in laymen's business." [93] That sentiment was a dogma on which, after the Restoration, both Cavalier and Roundhead could agree. It inevitably reacted, not only upon the practical powers of the clergy, which in any case had long been feeble, but on the whole conception of religion which regarded it as involving the control of economic self-interest by what Laud had called "the body of the Church." The works of Sanderson and of Jeremy Taylor, continuing an earlier tradition, reasserted with force and eloquence the view that the Christian is bound by his faith to a rule of life which finds expression in equity in bargaining and in works of mercy to his neighbors. [94] But the conception that the Church possessed, of its own authority, an independent standard of social values, which it could apply as a criterion to the practical affairs of the economic world, grew steadily weaker. The result, neither immediate nor intended, but inevitable, was the tacit denial of spiritual significance in the transactions of business and in the relations of organized society. Repudiating the right of religion to advance any social theory distinctively its own, that attitude became itself the most tyrannical and paralyzing of theories. It may be called Indifferentism.

The change had begun before the Civil War. It was completed with the Restoration, and, still more, with the Revolution. In the eighteenth century it is almost superfluous to examine the teaching of the Church of England as to social ethics. For it brings no distinctive contribution, and, except by a few eccentrics, the very conception of the Church as an independent moral authority, whose standards may be in sharp antithesis to social conventions, has been abandoned.

An institution which possesses no philosophy of its own inevitably accepts that which happens to be fashionable. What set the tone of social thought in the eighteenth century was partly the new Political Arithmetic, which had come to maturity at the Restoration, and which, as was to be expected in the first great age of English natural science—the age of Newton, of Halley, and of the Royal Society—drew its inspiration, not from religion or morals, but from mathematics and physics. It was still more the political theory associated with the name of Locke, but popularized and debased by a hundred imitators. Society is not a community of classes with varying functions, united to each other by mutual obligations arising from their relation to a common end. It is a joint-stock company rather than an organism, and the liabilities of the shareholders are strictly limited. They enter it in order to insure the rights already vested in them by the immutable laws of nature. The State, a matter of convenience, not of supernatural sanctions, exists for the protection of those rights, and fulfills its object in so far as, by maintaining contractual freedom, it secures full scope for their unfettered exercise.

The most important of such rights are property rights, and property rights attach mainly, though not, of course, exclusively, to the higher orders of men, who hold the tangible, material "stock" of society. Those who do not subscribe to the company have no legal claim to a share in the profits, though they have a moral claim on the charity of their superiors. Hence the curious phraseology which treats almost all below the nobility, gentry and freeholders as "the poor"—and the poor, it is well known, are of two kinds, "the industrious poor," who work for their betters, and "the idle poor," who work for themselves. Hence the unending discussions as to whether "the laboring poor" are to be classed among the "productive" or "unproductive" classes —whether they are, or are not, really worth their keep. Hence the indignant repudiation of the suggestion that any substantial amelioration of their lot could be effected by any kind of public policy. "It would be easier, where property was well secured, to live without money than without poor, . . . who, as they ought to be kept from starving, so they should receive nothing worth saving"; the poor "have nothing to stir them up to be serviceable but their wants, which it is prudence to relieve, but folly to cure"; "to make society happy, it is necessary that great numbers should be wretched as well as poor." [95] Such sentences from a work printed in 1714 are not typical. But they are straws which show how the wind is blowing.

In such an atmosphere temperatures were naturally low and equable, and enthusiasm, if not a lapse in morals, was an intellectual solecism and an error in taste. Religious thought was not immune from the same influence. It was not merely that the Church, which, as much as the State, was the heir of the Revolution settlement, reproduced the temper of an aristocratic society, as it reproduced its class organization and economic inequalities, and was disposed too often to idealize as a virtue that habit of mean subservience to wealth and social position, which, after more than half a century of political democracy, is still the characteristic and odious vice of Englishmen. Not less significant was the fact that, apart from certain groups and certain questions, it accepted the prevalent social philosophy and adapted its teaching to it. The age in which political theory was cast in the mould of religion had yielded to one in which religious thought was no longer an imperious master, but a docile pupil. Conspicuous exceptions like Law, who reasserted with matchless power the idea that Christianity implies a distinctive way of life, or protests like Wesley's sermon on

*The Use of Money,* merely heighten the impression of a
general acquiescence in the conventional ethics.   The prevalent religious thought might not unfairly be described as
morality tempered by prudence, and softened on occasion by
a rather sentimental compassion for inferiors.   It was the
natural counterpart of a social philosophy which repudiated
teleology, and which substituted the analogy of a self-regu-
lating mechanism, moved by the weights and pulleys of economic motives, for the theory which had regarded society
as an organism composed of different classes united by their
common subordination to a spiritual purpose.

Such an attitude, with its emphasis on the economic harmony of apparently conflicting interests, left small scope
for moral casuistry.   The materials for the reformer were,
indeed, abundant enough.   The phenomena of early commercial capitalism—consider only the orgy of financial immorality which culminated in 1720—were of a kind which
might have been expected to shock even the not over-sensi-
tive conscience of the eighteenth century.   Two centuries
before, the Fuggers had been denounced by preachers and
theologians; and, compared with the men who engineered
the South Sea Bubble, the Fuggers had been innocents.   In
reality, religious opinion was quite unmoved by the spec-
tacle.   The traditional scheme of social ethics had been
worked out in a simpler age; in the commercial England
of banking, and shipping, and joint-stock enterprise, it
seemed, and was called, a Gothic superstition.   From the
Restoration onward it was quietly dropped.   The usurer
and engrosser disappear from episcopal charges.   In the
popular manual called *The Whole Duty of Man,*[96] first published in 1658, and widely read during the following century, extortion and oppression still figure as sins, but the
attempt to define what they are is frankly abandoned.   If
preachers have not yet overtly identified themselves with
the view of the natural man, expressed by an eighteenth-

century writer in the words, "trade is one thing and religion is another," they imply a not very different conclusion by their silence as to the possibility of collisions between them. The characteristic doctrine was one, in fact, which left little room for religious teaching as to economic morality, because it anticipated the theory, later epitomized by Adam Smith in his famous reference to the invisible hand, which saw in economic self-interest the operation of a providential plan. "National commerce, good morals and good government," wrote Dean Tucker, of whom Warburton unkindly said that religion was his trade, and trade his religion, " are but part of one general scheme, in the designs of Providence."

Naturally, on such a view, it was unnecessary for the Church to insist on commercial morality, since sound morality coincided with commercial wisdom. The existing order, except in so far as the short-sighted enactments of Governments interfered with it, was the natural order, and the order established by nature was the order established by God. Most educated men, in the middle of the century, would have found their philosophy expressed in the lines of Pope:

> Thus God and Nature formed the general frame,
> And bade self-love and social be the same.

Naturally, again, such an attitude precluded a critical examination of institutions, and left as the sphere of Christian charity only those parts of life which could be reserved for philanthropy, precisely because they fell outside that larger area of normal human relations, in which the promptings of self-interest provided an all-sufficient motive and rule of conduct. It was, therefore, in the sphere of providing succor for the non-combatants and for the wounded, not in inspiring the main army, that the social work of the Church was conceived to lie. Its characteristic expressions in the eighteenth century were the relief of the poor, the care of the sick, and the establishment of schools. In spite of the genuine, if somewhat unctuous, solicitude for the spiritual welfare of the poorer classes, which inspired the Evangelical revival, religion abandoned the fundamental brain-work of criticism and construction to the rationalist and the humanitarian.

Surprise has sometimes been expressed that the Church should not have been more effective in giving inspiration and guidance during the immense economic reorganization to which tradition has assigned the not very felicitous name of the "Industrial Revolution." It did not give it, because it did not possess it. There were, no doubt, special conditions to account for its silence—mere ignorance and inefficiency, the supposed teachings of political economy, and, after 1790, the terror of all humanitarian movements inspired by France. But the explanation of its attitude is to be sought, less in the peculiar circumstances of the moment, than in the prevalence of a temper which accepted the established order of class relations as needing no vindication before any higher tribunal, and which made religion, not its critic or its accuser, but its anodyne, its apologist, and its drudge. It was not that there was any relapse into abnormal inhumanity. It was that the very idea that the Church possessed an independent standard of values, to which social institutions were amenable, had been abandoned. The surrender had been made long before the battle began. The spiritual blindness which made possible the general acquiescence in the horrors of the early factory system was, not a novelty, but the habit of a century.

# CHAPTER IV

## THE PURITAN MOVEMENT

"And the Lorde was with Joseph, and he was a luckie felowe."
*Genesis xxxix. 2 (Tyndale's translation).*

# CHAPTER IV

## THE PURITAN MOVEMENT

By the end of the sixteenth century the divorce between religious theory and economic realities had long been evident. But in the meantime, within the bosom of religious theory itself, a new system of ideas was being matured, which was destined to revolutionize all traditional values, and to turn on the whole field of social obligations a new and penetrating light. On a world heaving with expanding energies, and on a Church uncertain of itself, rose, after two generations of premonitory mutterings, the tremendous storm of the Puritan movement. The forest bent; the oaks snapped; the dry leaves were driven before a gale, neither all of winter nor all of spring, but violent and life-giving, pitiless and tender, sounding strange notes of yearning and contrition, as of voices wrung from a people dwelling in Meshec, which signifies Prolonging, in Kedar, which signifies Blackness; while amid the blare of trumpets, and the clash of arms, and the rending of the carved work of the Temple, humble to God and haughty to man, the soldier-saints swept over battlefield and scaffold their garments rolled in blood.

In the great silence which fell when the Titans had turned to dust, in the Augustan calm of the eighteenth century, a voice was heard to observe that religious liberty was a considerable advantage, regarded "merely in a commercial view."[1] A new world, it was evident, had arisen. And this new world, born of the vision of the mystic, the passion of the prophet, the sweat and agony of heroes famous and unknown, as well as of mundane ambitions and commonplace cupidities, was one in which, since "Thorough" was no more, since property was secure, and contracts inviolable, and the executive tamed, the judicious investments of business men were likely to yield a profitable return. So the epitaph, which crowns the life of what is called success, mocks the dreams in which youth hungered, not for success, but for the glorious failure of the martyr or the saint.

## I. PURITANISM AND SOCIETY

The principal streams which descended in England from the teaching of Calvin were three—Presbyterianism, Congregationalism, and a doctrine of the nature of God and man, which, if common to both, was more widely diffused, more pervasive and more potent than either. Of these three off-shoots from the parent stem, the first and eldest, which had made some stir under Elizabeth, and which it was hoped, with judicious watering from the Scotch, might grow into a State Church, was to produce a credal statement carved in bronze, but was to strike, at least in its original guise, but slender roots. The second, with its insistence on the right of every Church to organize itself, and on the freedom of all Churches from the interference of the State, was to leave, alike in the Old World and in the New, an imperishable legacy of civil and religious liberty. The third was Puritanism. Straitened to no single sect, and represented in the Anglican Church hardly, if at all, less fully than in those which afterwards separated from it, it determined, not only conceptions of theology and church government, but political aspirations, business relations, family life and the *minutiæ* of personal behavior.

The growth, triumph and transformation of the Puritan spirit was the most fundamental movement of the seventeenth century. Puritanism, not the Tudor secession from Rome, was the true English Reformation, and it is from its struggle against the old order that an England which is unmistakably modern emerges.  But, immense as were its accomplishments on the high stage of public affairs, its achievements in that inner world, of which politics are but the squalid scaffolding, were mightier still.  Like an iceberg, which can awe the traveller by its towering majesty only because sustained by a vaster mass which escapes his eye, the revolution which Puritanism wrought in Church and State was less than that which it worked in men's souls, and the watchwords which it thundered, amid the hum of Parliaments and the roar of battles, had been learned in the lonely nights, when Jacob wrestled with the angel of the Lord to wring a blessing before he fled.

> We do it wrong, being so majestical
> To offer it the show of violence.

In the mysticism of Bunyan and Fox, in the brooding melancholy and glowing energy of Cromwell, in the victorious tranquillity of Milton, "unshaken, unseduced, unterrified," amid a world of self-seekers and apostates, there are depths of light and darkness which posterity can observe with reverence or with horror, but which its small fathom-line cannot plumb.

There are types of character which are like a prism, whose various and brilliant colors are but broken reflections of a single ray of concentrated light.  If the inward and spiritual grace of Puritanism eludes the historian, its outward and visible signs meet him at every turn, and not less in market-place and counting-house and camp than in the student's chamber and the gathering of the elect for prayer. For to the Puritan, a contemner of the vain shows of sacramentalism, mundane toil becomes itself a kind of sacrament.  Like a man who strives by unresting activity to exorcise a haunting demon, the Puritan, in the effort to save his own soul, sets in motion every force in heaven above or in the earth beneath.  By the mere energy of his expanding spirit, he remakes, not only his own character and habits and way of life, but family and church, industry and city, political institutions and social order.  Conscious that he is but a stranger and pilgrim, hurrying from this transitory life to a life to come, he turns with almost physical horror from the vanities which lull into an awful indifference souls dwelling on the borders of eternity, to pore with anguish of spirit on the grand facts, God, the soul, salvation and damnation.  "It made the world seem to me," said a Puritan of his conversion, "as a carkass that had neither life nor loveliness.  And it destroyed those ambitious desires after literate fame, which was the sin of my childhood. . . . It set me upon that method of my studies which since then I have found the benefit of. . . . It caused me first to seek God's Kingdom and his Righteousness, and most to mind the One thing needful, and to determine first of my Ultimate End." [2]

Overwhelmed by a sense of his "Ultimate End," the Puritan cannot rest, nevertheless, in reflection upon it.  The contemplation of God, which the greatest of the Schoolmen described as the supreme blessedness, is a blessedness too great for sinners, who must not only contemplate God, but glorify him by their work in a world given over to the powers of darkness.  "The way to the Celestial City lies just through this town, where this lusty fair is kept; and he that will go to the City, and yet not go through this town, must needs go out of the world." [3]  For that awful journey, girt with precipices and beset with fiends, he sheds every encumbrance, and arms himself with every weapon.  Amusements, books, even intercourse with friends, must, if need be, be cast aside; for it is better to enter into eternal life halt and maimed than having two eyes to be cast into eternal fire. He scours the country, like Baxter and Fox, to find one who may speak the word of life to his soul.  He seeks from his ministers, not absolution, but instruction, exhortation and warning. Prophesyings—that most revealing episode in early Puritanism—were the cry of a famished generation for enlightenment, for education, for a religion of the intellect; and it was because much "preaching breeds faction, but much praying causes devotion"[4] that the powers of this world raised their parchment shutters to stem the gale that blew from the Puritan pulpit. He disciplines, rationalizes, systematizes, his life; "method" was a Puritan catchword a century before the world had heard of Methodists. He makes his very business a travail of the spirit, for that too is the Lord's vineyard, in which he is called to labor.

Feeling in him that which "maketh him more fearful of displeasing God than all the world,"[5] he is a natural republican, for there is none on earth that he can own as master. If powers and principalities will hear and obey, well; if not, they must be ground into dust, that on their ruins the elect may build the Kingdom of Christ. And, in the end, all these—prayer, and toil, and discipline, mastery of self and mastery of others, wounds, and death—may be too little for the salvation of a single soul. "Then I saw that there was a way to Hell even from the Gates of Heaven, as well as from the City of Destruction"[6]—those dreadful words haunt him as he nears his end. Sometimes they break his heart. More often, for grace abounds even to the chief of sinners, they nerve his will. For it is will—will organized and disciplined and inspired, will quiescent in rapt adoration or straining in violent energy, but always will—which is the essence of Puritanism, and for the intensification and organization of will every instrument in that tremendous arsenal of religious fervour is mobilized. The Puritan is like a steel spring compressed by an inner force, which shatters every obstacle by its rebound. Sometimes the strain is too tense, and, when its imprisoned energy is released, it shatters itself.

The spirit bloweth where it listeth, and men of every so-

cial grade had felt their hearts lifted by its breath, from
aristocrats and country gentlemen to weavers who, "as they
stand in their loom, can set a book before them or edifie
one another." [7]  But, if religious zeal and moral enthusiasm are not straitened by the vulgar categories of class
and income, experience proves, nevertheless, that there are
certain kinds of environment in which they burn more
bravely than in others, and that, as man is both spirit and
body, so different types of religious experience correspond
to the varying needs of different social and economic *milieux*.
To contemporaries the chosen seat of the Puritan spirit
seemed to be those classes in society which combined economic independence, education and a certain decent pride in
their status, revealed at once in a determination to live their
own lives, without truckling to earthly superiors, and in a
somewhat arrogant contempt for those who, either through
weakness of character or through economic helplessness, were
less resolute, less vigorous and masterful, than themselves.
Such, where the feudal spirit had been weakened by contact with town life and new intellectual currents, were some
of the gentry.   Such, conspicuously, were the yeomen,
"mounted on a high spirit, as being slaves to none," [8] especially in the freeholding counties of the east.   Such, above
all, were the trading classes of the towns, and of those rural
districts which had been partially industrialized by the de-
centralization of the textile and iron industries.

"The King's cause and party," wrote one who described
the situation in Bristol in 1645, "were favored by two ex-
tremes in that city; the one, the wealthy and powerful men,
the other, of the basest and lowest sort; but disgusted by
the middle rank, the true and best citizens." [9]   That it was
everywhere these classes who were the standard-bearers of
Puritanism is suggested by Professor Usher's statistical estimate of the distribution of Puritan ministers in the first
decade of the seventeenth century, which shows that, of 281

ministers whose names are known, 35 belonged to London
and Middlesex, 96 to the three manufacturing counties of
Norfolk, Suffolk and Essex, 29 to Northamptonshire, 17
to Lancashire, and only 104 to the whole of the rest of the
country.[10] The phenomenon was so striking as to evoke the
comments of contemporaries absorbed in matters of pro-
founder spiritual import than sociological generalization.
"Most of the tenants of these gentlemen," wrote Baxter,
"and also most of the poorest of the people, whom the other
called the Rabble, did follow the gentry, and were for the
King. On the Parliament's side were (besides themselves)
the smaller part (as some thought) of the gentry in most
of the counties, and freeholders, and the middle sort of men;
especially in those corporations and counties which depend
on cloathing and such manufactures." He explained the
fact by the liberalizing effect of constant correspondence
with the greater centers of trade, and cited the example of
France, where it was "the merchants and middle sort of
men that were Protestants." [11]

The most conspicuous example was, of course, London,
which had financed the Parliamentary forces, and which
continued down to the Revolution to be *par excellence* "the
rebellious city," returning four Dissenters to the Royalist
Parliament of 1661, sending its mayor and aldermen to accompany Lord Russell when he carried the Exclusion Bill
from the Commons to the Lords, patronizing Presbyterian
ministers long after Presbyterianism was proscribed, nurs-
ing the Whig Party, which stood for tolerance, and shel-
tering the Whig leaders against the storm which
broke in 1681. But almost everywhere the same fact was
to be observed. The growth of Puritanism, wrote a hostile
critic, was "by meanes of the City of London (the nest and
seminary of the seditious faction) and by reason of its uni-
versall trade throughout the kingdome, with its commodities
conveying and deriving this civill contagion to all our cities

and corporations, and thereby poysoning whole counties." [12]
In Lancashire, the clothing towns—"the Genevas of Lancashire"—rose like Puritan islands from the surrounding
sea of Roman Catholicism.  In Yorkshire, Bradford, Leeds
and Halifax; in the midlands, Birmingham and Leicester; in
the west, Gloucester, Taunton and Exeter, the capital of
the west of England textile industry, were all centers of
Puritanism.

The identification of the industrial and commercial classes
with religious radicalism was, indeed, a constant theme of
Anglicans and Royalists, who found in the vices of each an
additional reason for distrusting both.  Clarendon commented bitterly on the "factious humor which possessed
most corporations, and the pride of their wealth"; [13] and,
after the Civil War, both the politics and the religion of
the boroughs were suspect for a generation.  The bishop
of Oxford warned Charles II's Government against showing them any favor, on the ground that "trading combinations" were "so many nests of faction and sedition," and
that "our late miserable distractions" were "chiefly hatched
in the shops of tradesmen." [14]  Pepys commented dryly on
the black looks which met the Anglican clergy as they returned to their City churches.  It was even alleged that the
courtiers hailed with glee the fire of London, as a providential instrument for crippling the center of disaffection.[15]

When, after 1660, Political Arithmetic became the fashion, its practitioners were moved by the experience of the
last half-century and by the example of Holland—the economic schoolmaster of seventeenth-century Europe—to inquire, in the manner of any modern sociologist, into the relations between economic progress and other aspects of the
national genius.  Cool, dispassionate, very weary of the
drum ecclesiastic, they confirmed, not without some notes of
gentle irony, the diagnosis of bishop and presbyterian, but
deduced from it different conclusions.  The question which

gave a topical point to their analysis was the rising issue of religious tolerance.  Serenely indifferent to its spiritual significance, they found a practical reason for applauding it in the fact that the classes who were in the van of the Puritan movement, and in whom the Clarendon Code found its most prominent victims, were also those who led commercial and industrial enterprise.  The explanation, they thought, was simple.  A society of peasants could be homogeneous in its religion, as it was already homogeneous in the simple uniformity of its economic arrangements.  A many-sided business community could escape constant friction and obstruction only if it were free to absorb elements drawn from a multitude of different sources, and if each of those elements were free to pursue its own way of life, and—in that age the same thing—to practice its own religion.

Englishmen, as Defoe remarked, improved everything and invented nothing, and English economic organization had long been elastic enough to swallow Flemish weavers flying from Alva, and Huguenots driven from France.  But the traditional ecclesiastical system was not equally accommodating.  It found not only the alien refugee, but its home-bred sectaries, indigestible.  Laud, reversing the policy of Elizabethan Privy Councils, which characteristically thought diversity of trades more important than unity of religion, had harassed the settlements of foreign artisans at Maidstone, Sandwich and Canterbury,[16] and the problem recurred in every attempt to enforce conformity down to 1689.  "The gaols were crowded with the most substantial tradesmen and inhabitants, the clothiers were forced from their houses, and thousands of workmen and women whom they employed set to starving."[17]  The Whig indictment of the disastrous effects of Tory policy recalls the picture drawn by French *intendants* of the widespread distress which followed the revocation of the Edict of Nantes.[18]

When the collision between economic interests and the

policy of compulsory conformity was so flagrant, it is not surprising that the economists of the age should have enunciated the healing principle that persecution was incompatible with prosperity, since it was on the pioneers of economic progress that persecution principally fell. "Every law of this nature," wrote the author of a pamphlet on the subject, is not only "expressly against the very principles and rules of the Gospel of Christ," but is also "destructive to the trade and well-being of our nation by oppressing and driving away the most industrious working hands, and depopulating, and thereby impoverishes our country, which is capable of employing ten times the number of people we now have." [19]

Temple, in his calm and lucid study of the United Netherlands, found one reason of their success in the fact that, Roman Catholicism excepted, every man might practise what religion he pleased.[20] De la Court, whose striking book passed under the name of John de Witt, said the same.[21] Petty, after pointing out that in England the most thriving towns were those where there was most nonconformity, cited the evidence, not only of Europe, but of India and the Ottoman Empire, to prove that, while economic progress is compatible with any religion, the class which is its vehicle will always consist of the heterodox minority, who "profess opinions different from what are publicly established." [22] "There is a kind of natural unaptness," wrote a pamphleteer in 1671, "in the Popish religion to business, whereas on the contrary among the Reformed, the greater their zeal, the greater their inclination to trade and industry, as holding idleness unlawful. . . . The domestic interest of England lieth in the advancement of trade by removing all obstructions both in city and country, and providing such laws as may help it, and make it most easy, especially in giving liberty of conscience to all Protestant Nonconformists, and denying it to Papists." [23]

If the economists applauded tolerance because it was good for trade, the Tory distrust of the commercial classes was aggravated by the fact that it was they who were most vocal in the demand for tolerance. Swift denounced, as part of the same odious creed, the maxim that "religion ought to make no distinction between Protestants" and the policy "of preferring, on all occasions, the monied interests before the landed." [24] Even later in the eighteenth century, the stale gibe of "the Presbyterians, the Bank and the other corporations" still figured in the pamphlets of the statesman whom Lord Morley describes as the prince of political charlatans, Bolingbroke.[25]

"The middle ranks," "the middle class of men," "the middle sort"—such social strata included, of course, the widest variety of economic interest and personal position. But in the formative period of Puritanism, before the Civil War, two causes prevented the phrase from being merely the vapid substitute for thought which it is today. In the first place, outside certain exceptional industries and districts, there was little large-scale production and no massed proletariat of propertyless wage-earners. As a result, the typical workman was still normally a small master, who continued himself to work at the loom or at the forge, and whose position was that described in Baxter's Kidderminster, where "there were none of the tradesmen very rich . . . the magistrates of the town were few of them worth £40 *per annum,* and most not half so much; three or four of the richest thriving masters of the trade got but about £500 to £600 in twenty years, and it may be lost £100 of it at once by an ill debtor." [26] Differing in wealth from the prosperous merchant or clothier, such men resembled them in economic and social habits, and the distinction between them was one of degree, not of kind. In the world of industry vertical divisions between district and district still cut deeper than horizontal fissures between class and class.

The number of those who could reasonably be described as independent, since they owned their own tools and controlled their own businesses, formed a far larger proportion of the population than is the case in capitalist societies.

The second fact was even more decisive. The business classes, as a power in the State, were still sufficiently young to be conscious of themselves as something like a separate order, with an outlook on religion and politics peculiarly their own, distinguished, not merely by birth and breeding, but by their social habits, their business discipline, the whole bracing atmosphere of their moral life, from a Court which they believed to be godless and an aristocracy which they knew to be spendthrift. The estrangement—for it was no more—was of shorter duration in England than in any other European country, except Switzerland and Holland. By the latter part of the seventeenth century, partly as a result of the common struggles which made the Revolution, still more perhaps through the redistribution of wealth by commerce and finance, the former rivals were on the way to be compounded in the gilded clay of a plutocracy embracing both. The landed gentry were increasingly sending their sons into business; "the tradesman meek and much a liar" looked forward, as a matter of course, to buying an estate from a bankrupt noble. Georgian England was to astonish foreign observers, like Voltaire and Montesquieu, as the Paradise of the *bourgeoisie,* in which the prosperous merchant shouldered easily aside the impoverished bearers of aristocratic names.[27]

That consummation, however, was subsequent to the great divide of the Civil War, and, in the main, to the tamer glories of the Revolution. In the germinating period of Puritanism, the commercial classes, though powerful, were not yet the dominant force which a century later they were to become. They could look back on a not distant past, in which their swift rise to prosperity had been regarded with suspicion, as the emergence of an alien interest, which applied sordid means to the pursuit of anti-social ends—an interest for which in a well-ordered commonwealth there was little room, and which had been rapped on the knuckles by conservative statesmen. They lived in a present, where a Government, at once interfering, inefficient and extravagant, cultivated, with an intolerable iteration of grandiloquent principles, every shift and artifice most repugnant to the sober prudence of plain-dealing men. The less reputable courtiers and the more feather-pated provincial gentry, while courting them to raise a mortgage or renew a loan, reviled them as *parvenus,* usurers and blood-suckers. Even in the latter part of the seventeenth century, the influence of the *rentier* and of the financier still continued to cause apprehension and jealousy, both for political and for economic reasons. "By this single stratagem," wrote an indignant pamphleteer of the Puritan capitalists who specialized in money-lending, "they avoyd all contributions of tithes and taxes to the King, Church, Poor (a soverain cordial to tender consciences) ; they decline all services and offices of burthen incident to visible estates; they escape all oaths and ties of publick allegiance or private fealty. . . . They enjoy both the secular applause of prudent conduct, and withal the spiritual comfort of thriving easily and devoutly . . . leaving their adversaries the censures of improvidence, together with the misery of decay. They keep many of the nobility and gentry in perfect vassalage (as their poor copyholders), which eclipses honour, enervates justice and ofttimes protects them in their boldest conceptions. By engrossing cash and credit, they in effect give the price to land and law to markets. By commanding ready money, they likewise command such offices as they widely affect . . . they feather and enlarge their own nests, the corporations." [28]

Such lamentations, the protest of senatorial dignity against equestrian upstarts or of the *noblesse* against the *roturier,* were natural in a conservative aristocracy, which for a century had felt authority and prestige slipping from its grasp, and which could only maintain its hold on them by resigning itself, as ultimately it did, to sharing them with its rival. In return, the business world, which had its own religious and political ideology, steadily gathered the realities of power into its own hands; asked with a sneer, "how would merchants thrive if gentlemen would not be unthriftes"; [29] and vented the indignant contempt felt by an energetic, successful and, according to its lights, not too unscrupulous, generation for a class of *fainéants,* unversed in the new learning of the City and incompetent to the verge of immorality in the management of business affairs. Their triumphs in the past, their strength in the present, their confidence in the future, their faith in themselves, and their difference from their feebler neighbours—a difference as of an iron wedge in a lump of clay—made them, to use a modern phrase, class-conscious. Like the modern proletarian, who feels that, whatever his personal misery and his present disappointments, the Cause is rolled forward to victory by the irresistible force of an inevitable evolution, the Puritan *bourgeoisie* knew that against the chosen people the gates of hell could not prevail. The Lord prospered their doings.

There is a magic mirror in which each order and organ of society, as the consciousness of its character and destiny dawns upon it, looks for a moment, before the dust of conflict or the glamour of success obscures its vision. In that enchanted glass, it sees its own lineaments reflected with ravishing allurements; for what it sees is not what it is, but what in the eyes of mankind and of its own heart it would be. The feudal *noblesse* had looked, and had caught a glimpse of a world of fealty and chivalry and honor. The monarchy looked, or Laud and Strafford looked for it; they saw a nation drinking the blessings of material prosperity and spiritual edification from the cornucopia of a sage and paternal monarchy—a nation "fortified and adorned . . . the country rich . . . the Church flourishing . . . trade increased to that degree that we were the exchange of Christendom . . . all foreign merchants looking upon nothing as their own but what they laid up in the warehouses of this Kingdom."[30]  In a far-off day the craftsman and laborer were to look, and see a band of comrades, where fellowship should be known for life and lack of fellowship for death. For the middle classes of the early seventeenth century, rising but not yet triumphant, that enchanted mirror was Puritanism.  What it showed was a picture grave to sternness, yet not untouched with a sober exaltation—an earnest, zealous, godly generation, scorning delights, punctual in labor, constant in prayer, thrifty and thriving, filled with a decent pride in themselves and their calling, assured that strenuous toil is acceptable to Heaven, a people like those Dutch Calvinists whose economic triumphs were as famous as their iron Protestantism—"thinking, sober, and patient men, and such as believe that labor and industry is their duty towards God."[31]  Then an air stirred and the glass was dimmed.  It was long before any questioned it again.

II.  A GODLY DISCIPLINE *versus* THE RELIGION OF TRADE

Puritanism was the schoolmaster of the English middle classes.  It heightened their virtues, sanctified, without eradicating, their convenient vices, and gave them an inexpugnable assurance that, behind virtues and vices alike, stood the majestic and inexorable laws of an omnipotent Providence, without whose foreknowledge not a hammer could beat upon the forge, not a figure could be added to the ledger.  But it is a strange school which does not teach more than one lesson, and the social reactions of Puritanism, trenchant, permanent and profound, are not to be summarized in the simple formula that it fostered individualism. Weber, in a celebrated essay, expounded the thesis that Calvinism, in its English version, was the parent of capitalism, and Troeltsch, Schulze-Gaevernitz and Cunningham have lent to the same interpretation the weight of their considerable authority.[32] But the heart of man holds mysteries of contradiction which live in vigorous incompatibility together. When the shriveled tissues lie in our hand, the spiritual bond still eludes us.

In every human soul there is a socialist and an individualist, an authoritarian and a fanatic for liberty, as in each there is a Catholic and a Protestant. The same is true of the mass movements in which men marshal themselves for common action. There was in Puritanism an element which was conservative and traditionalist, and an element which was revolutionary; a collectivism which grasped at an iron discipline, and an individualism which spurned the savorless mess of human ordinances; a sober prudence which would garner the fruits of this world, and a divine recklessness which would make all things new. For long nourished together, their discords concealed, in the furnace of the Civil War they fell apart, and Presbyterian and Independent, aristocrat and Leveller, politician and merchant and utopian, gazed with bewildered eyes on the strange monsters with whom they had walked as friends. Then the splendors and illusions vanished; the force of common things prevailed; the metal cooled in the mould; and the Puritan spirit, shorn of its splendors and its illusions, settled finally into its decent bed of equable respectability. But each element in its social philosophy had once been as vital as the other, and the battle was fought, not between a Puritanism solid for one view and a State committed to another, but between rival tendencies in the soul of Puritanism itself. The problem is to grasp their connection, and to understand the reasons which caused this to wax and that to wane.

"The triumph of Puritanism," it has been said, "swept away all traces of any restriction or guidance in the employment of money." [33]   That it swept away the restrictions imposed by the existing machinery is true; neither ecclesiastical courts, nor High Commission, nor Star Chamber, could function after 1640. But, if it broke the discipline of the Church of Laud and the State of Strafford, it did so but as a step towards erecting a more rigorous discipline of its own. It would have been scandalized by economic individualism as much as by religious tolerance, and the broad outlines of its scheme of organization favored unrestricted liberty in matters of business as little as in the things of the spirit. To the Puritan of any period in the century between the accession of Elizabeth and the Civil War, the suggestion that he was the friend of economic or social license would have seemed as wildly inappropriate as it would have appeared to most of his critics, who taunted him, except in the single matter of usury, with an intolerable meticulousness.

A godly discipline was, indeed, the very ark of the Puritan covenant. Delivered in thunder to the Moses of Geneva, its vital necessity had been the theme of the Joshuas of Scotland, England and France. Knox produced a Scottish edition of it; Cartwright, Travers and Udall composed treatises expounding it. Bancroft exposed its perils for the established ecclesiastical order.[34]   The word "discipline" implied essentially "a directory of Church government," established in order that "the wicked may be corrected with ecclesiastical censures, according to the quality of the fault"; [35] and the proceedings of Puritan *classes* in the sixteenth century show that the conception of a rule of life, to be enforced by the pressure of the common conscience, and in the last resort by spiritual penalties, was a vital part of their system. When, at the beginning of Elizabeth's reign,

the sectaries in London described their objects as not merely
the "free and pure" preaching of the Gospel, nor the pure
ministration of the sacraments, but "to have, not the fylthye
cannon lawe, but disciplyne onelye and altogether agreeable
to the same heavenlye and Allmightye word of our good
Lorde Jesus Chryste," [36] the antithesis suggests that something more than verbal instruction is intended.  Bancroft
noted that it was the practice, when a sin was committed
by one of the faithful, for the elders to apply first admon-
ishment and then excommunication.  The minute-book of
one of the few *classes* whose records survive confirms his
statement.[37]

All this early movement had almost flickered out before
the end of the sixteenth century.  But the conception lay
at the very root of Presbyterianism, and it reëmerged in
the system of church government which the supercilious
Scotch Commissioners at the Westminster Assembly steered
to inconclusive victory, between Erastians on the right and
Independents on the left.  The destruction of the Court of
High Commission, of the temporal jurisdiction of all persons in Holy Orders, and finally, with the abolition of
episcopacy, of the ecclesiastical courts themselves, left a
vacuum.  "Mr. Henderson," wrote the insufferable Baillie,
"has ready now a short treatise, much called for, of our
church discipline." [38]  In June 1646 an unenthusiastic Parliament accepted the ordinance which, after a three years'
debate of intolerable tedium, emerged from the Assembly's
Committee on the Discipline and Government of the Church,
and which provided for the suspension by the elders of
persons guilty of scandalous offences.  Detested by the Independents and cold-shouldered by Parliament, which had
no intention of admitting the divine right of presbyteries,
the system never took deep root, and in London, at least,
there appears to be no evidence of any exercise of jurisdiction by elders or *classes*.  In parts of Lancashire, on the

other hand, it seems to have been actively at work, down, at any rate, to 1649. The change in the political situation, in particular the triumph of the army, prevented it, Mr. Shaw thinks, from functioning longer.[39]

"Discipline" included all questions of moral conduct, and of these, in an age when a great mass of economic relations were not the almost automatic reactions of an impersonal mechanism, but a matter of human kindliness or meanness between neighbors in village or borough, economic conduct was naturally part. Calvin and Beza, perpetuating with a new intensity the medieval idea of a Church-civilization, had sought to make Geneva a pattern, not only of doctrinal purity, but of social righteousness and commercial morality. Those who had drunk from their spring continued, in even less promising environments, the same tradition. Bucer, who wrote when something more fundamental than a politician's reformation seemed possible to enthusiasts with their eyes on Geneva, had urged the reconstruction of every side of the economic life of a society which was to be Church and State in one.[40] English Puritanism, while accepting after some hesitation Calvin's much qualified condonation of moderate interest, did not intend in other respects to countenance a laxity welcome only to worldlings. Knewstub appealed to the teaching of "that worthy instrument of God, Mr. Calvin," to prove that the habitual usurer ought to be "thrust out of the society of men." Smith embroidered the same theme. Baro, whose Puritanism lost him his professorship, denounced the "usual practice amongst rich men, and some of the greater sort, who by lending, or by giving out their money to usury, are wont to snare and oppress the poor and needier sort." Cartwright, the most famous leader of Elizabethan Puritanism, described usury as "a hainous offence against God and his Church," and laid down that the offender should be excluded from the sacraments until he satisfied the congregation of his penitence.[41]

The ideal of all was that expressed in the apostolic injunction to be content with a modest competence and to shun the allurements of riches. "Every Christian man is bound in conscience before God," wrote Stubbes, "to provide for his household and family, but yet so as his immoderate care surpasse not the bands, nor yet transcend the limits, of true Godlynes. . . . So farre from covetousnes and from immoderate care would the Lord have us, that we ought not this day to care for tomorrow, for (saith he) sufficient to the day is the travail of the same." [42]

The most influential work on social ethics written in the first half of the seventeenth century from the Puritan standpoint was Ames' *De Conscientia,* a manual of Christian conduct which was intended to supply the brethren with the practical guidance which had been offered in the Middle Ages by such works as *Dives et Pauper.* It became a standard authority, quoted again and again by subsequent writers. Forbidden to preach by the bishop of London, Ames spent more than twenty years in Holland, where he held a chair of theology at the University of Franeker, and his experience of social life in the country which was then the business capital of Europe makes the remorseless rigor of his social doctrine the more remarkable. He accepts, as in his day was inevitable, the impossibility of distinguishing between interest on capital invested in business, and interest on capital invested in land, since men put money indifferently into both, and, like Calvin, he denies that interest is forbidden in principle by Scripture or natural reason. But, like Calvin, he surrounds his indulgence with qualifications; he requires that no interest shall be charged on loans to the needy, and describes as the ideal investment for Christians one in which the lender shares risks with the borrower, and demands only "a fair share of the profits, according to the degree in which God has blessed him by whom the money is used." His teaching with regard to prices is not less conservative.

"To wish to buy cheap and to sell dear is common (as Augustine observes), but it is a common vice." Men must not sell above the maximum fixed by public authority, though they may sell below it, since it is fixed to protect the buyer; when there is no legal maximum, they must follow the market price and "the judgment of prudent and good men." They must not take advantage of the necessities of individual buyers, must not overpraise their wares, must not sell them dearer merely because they have cost them much to get.[43] Puritan utterances on the subject of enclosing were equally trenchant.[44]

Nor was such teaching merely the pious pedantry of the pulpit. It found some echo in contrite spirits; it left some imprint on the conduct of congregations. If D'Ewes was the unresisting victim of a more than ordinarily aggressive conscience, he was also a man of the world who played a not inconspicuous part in public affairs; and D'Ewes not only ascribed the fire which destroyed his father's house to the judgment of Heaven on ill-gotten gains, but expressly prescribed in his will that, in order to avoid the taint of the accursed thing, provision should be made for his daughters, not by investing his capital at a fixed—and therefore usurious—rate of interest, but by the purchase either of land or of annuities.[45] The *classis* which met at Dedham in the eighties of the sixteenth century was concerned partly with questions of ceremony, of church government, of the right use of Sunday, and with the weighty problems whether boys of sixteen might wear their hats in church, and by what marks one might detect a witch. But it discussed also what provision could be made to check vagrancy; advised the brethren to confine their dealings to "the godliest of that trade" (of cloth making); recommended the establishment in the township of a scheme of universal education, that of children of parents too poor to meet the cost being defrayed from collections made in church; and urged that each well-to-do householder should provide in his home for two (or, if less able, one) of his impoverished neighbors who "walke christianly and honestlie in their callinges." [46] In the ever-lengthening list of scandalous and notorious sins to be punished by exclusion from the sacrament, which was elaborated by the Westminster Assembly, a place was found, not only for drunkards, swearers, and blasphemers, worshippers and makers of images, senders or carriers of challenges, persons dancing, gaming, attending plays on the Lord's day, or resorting to witches, wizards, and fortune-tellers, but for the more vulgar vices of those who fell into extortion, barratry and bribery.[47] The *classis* of Bury in Lancashire (*quantum mutatus!*) took these economic lapses seriously. It decided in 1647, after considerable debate, that "usury is a scandalous sin, deserving suspention upon obstinacy." [48]

It was a moment when good men were agog to cast the money-changers from the temple and to make straight the way of the Lord. "God hath honnored you in callinge you to a place of power and trust, and hee expects that you should bee faithfull to that trust. You are postinge to the grave every day; you dwell uppon the borders of eternity; your breath is in your nostrells; therfore duble and treble your resolutions to bee zealous in a good thinge. . . . How dreadfull will a dieinge bed bee to a negligent magistrate! What is the reward of a slothfull servant? Is it not to bee punished with everlastinge destruction from the presence of the Lord?" [49] Such, in that singular age, was the language in which the mayor of Salisbury requested the justices of Wiltshire to close four public-houses. Apparently they closed them.

The attempt to crystallize social morality in an objective discipline was possible only in a theocracy; and, still eloquent in speech, theocracy had abdicated in fact, even before the sons of Belial returned to cut down its groves and lay waste its holy places. In an age when the right to dissent from the State Church was still not fully established, its defeat was fortunate, for it was the victory of tolerance. It meant, however, that the discipline of the Church gave place to the attempt to promote reform through the action of the State, which reached its height in the Barebones Parliament. Projects for law reform, marriage reform and financial reform, the reform of prisons and the relief of debtors, jostled each other on its committees; while outside it there were murmurs among radicals against social and economic privilege, which were not to be heard again till the days of the Chartists, and which to the conservative mind of Cromwell seemed to portend mere anarchy. The transition from the idea of a moral code enforced by the Church, which had been characteristic of early Calvinism, to the economic individualism of the later Puritan movement took place, in fact, by way of the democratic agitation of the Independents. Abhorring the whole mechanism of ecclesiastical discipline and compulsory conformity, they endeavored to achieve the same social and ethical ends by political action.

The change was momentous. If the English Social Democratic movement has any single source, that source is to be found in the New Model Army. But the conception implied in the attempt to formulate a scheme of economic ethics—the theory that every department of life falls beneath the same all-encompassing arch of religion—was too deeply rooted to be exorcised merely by political changes, or even by the more corroding march of economic development. Expelled from the world of fact, where it had always been a stranger and a sojourner, it survived in the world of ideas, and its champions in the last half of the century labored it the more, precisely because they knew that it must be conveyed to their audiences by teaching and preaching or not at all. Of those champions the most learned, the most practical, and the most persuasive was Richard Baxter.

How Baxter endeavored to give practical instruction to

his congregation at Kidderminster, he himself has told us. "Every Thursday evening my neighbours that were most desirous and had opportunity met at my house, and there one of them repeated the sermon, and afterwards they proposed what doubts any of them had about the sermon, or any other case of conscience, and I resolved their doubts." [50] Both in form and in matter, his *Christian Directory, or a Summ of Practical Theologie and Cases of Conscience* [51] is a remarkable book. It is, in essence, a Puritan *Summa Theologica* and *Summa Moralis* in one; its method of treatment descends directly from that of the medieval *Summæ,* and it is, perhaps, the last important English specimen of a famous *genus.* Its object, as Baxter explains in his introduction, is "the resolving of practical cases of conscience, and the reducing of theoretical knowledge into serious Christian practice." Divided into four parts, Ethics, Economics, Ecclesiastics, and Politics, it has as its purpose to establish the rules of a Christian casuistry, which may be sufficiently detailed and precise to afford practical guidance to the proper conduct of men in the different relations of life, as lawyer, physician, schoolmaster, soldier, master and servant, buyer and seller, landlord and tenant, lender and borrower, ruler and subject. Part of its material is derived from the treatment of similar questions by previous writers, both before and after the Reformation, and Baxter is conscious of continuing a great tradition. But it is, above all things, realistic, and its method lends plausibility to the suggestion that it originated in an attempt to answer practical questions put to its author by members of his congregation. Its aim is not to overwhelm by authority, but to convince by an appeal to the enlightened common sense of the Christian reader. It does not overlook, therefore, the practical facts of a world in which commerce is carried on by the East India Company in distant markets, trade is universally conducted on credit, the iron manufacture is a large-scale industry demanding abundant supplies of capital and offering a profitable opening to the judicious investor, and the relations of landlords and tenants have been thrown into confusion by the fire of London. Nor does it ignore the moral qualities for the cultivation of which an opportunity is offered by the life of business. It takes as its starting-point the commercial environment of the Restoration, and its teaching is designed for "Rome or London, not Fools' Paradise."

Baxter's acceptance of the realities of his age makes the content of his teaching the more impressive. The attempt to formulate a casuistry of economic conduct obviously implies that economic relations are to be regarded merely as one department of human behavior, for which each man is morally responsible, not as the result of an impersonal mechanism, to which ethical judgments are irrelevant. Baxter declines, therefore, to admit the convenient dualism, which exonerates the individual by representing his actions as the outcome of uncontrollable forces. The Christian, he insists, is committed by his faith to the acceptance of certain ethical standards, and these standards are as obligatory in the sphere of economic transactions as in any other province of human activity. To the conventional objection that religion has nothing to do with business—that "every man will get as much as he can have and that *caveat emptor* is the only security"—he answers bluntly that this way of dealing does not hold among Christians. Whatever the laxity of the law, the Christian is bound to consider first the golden rule and the public good. Naturally, therefore, he is debarred from making money at the expense of other persons, and certain profitable avenues of commerce are closed to him at the outset. "It is not lawful to take up or keep up any oppressing monopoly or trade, which tends to enrich you by the loss of the Commonwealth or of many."

But the Christian must not only eschew the obvious extortion practiced by the monopolist, the engrosser, the organizer of a corner or a combine. He must carry on his business in the spirit of one who is conducting a public service; he must order it for the advantage of his neighbor as much as, and, if his neighbor be poor, more than, for his own. He must not desire "to get another's goods or labour for less than it is worth." He must not secure a good price for his own wares "by extortion working upon men's ignorance, error, or necessity." When prices are fixed by law, he must strictly observe the legal maximum; when they are not, he must follow the price fixed by common estimation. If he finds a buyer who is willing to give more, he "must not make too great an advantage of his convenience or desire, but be glad that [he] can pleasure him upon equal, fair, and honest terms," for "it is a false rule of them that think their commodity is worth as much as any one will give." If the seller foresees that in the future prices are likely to fall, he must not make profit out of his neighbour's ignorance, but must tell him so. If he foresees that they will rise, he may hold his wares back, but only—a somewhat embarrassing exception—if it be not "to the hurt of the Commonwealth, as if . . . keeping it in be the cause of the dearth, and . . . bringing it forth would help to prevent it." If he is buying from the poor, "charity must be exercised as well as justice"; the buyer must pay the full price that the goods are worth to himself, and, rather than let the seller suffer because he cannot stand out for his price, should offer him a loan or persuade some one else to do so. In no case may a man doctor his wares in order to get for them a higher price than they are really worth, and in no case may he conceal any defects of quality; if he was so unlucky as to have bought an inferior article, he "may not repair [his] loss by doing as [he] was done by, . . . no more than [he] may cut another's purse because [his] was cut." Rivalry in trade, Baxter thinks, is inevitable. But the Christian must not snatch a good bargain "out of greedy covetousness, nor to the injury of the poor . . . nor . . . so as to disturb that due and civil order which should be among moderate men in trading." On the contrary, if "a covetous oppressor" offer a poor man less than his goods are worth, "it may be a duty to offer the poor man the worth of his commodity and save him from the oppressor."

The principles which should determine the contract between buyer and seller are applied equally to all other economic relations. Usury, in the sense of payment for a loan, is not in itself unlawful for Christians. But it becomes so, when the lender does not allow the borrower "such a proportion of the gain as his labour, hazard, or poverty doth require, but . . . will live at ease upon his labours"; or when, in spite of the borrower's misfortune, he rigorously exacts his pound of flesh; or when interest is demanded for a loan which charity would require to be free. Masters must discipline their servants for their good; but it is "an odious oppression and injustice to defraud a servant or labourer of his wages, yea, or to give him less than he deserveth." As the descendant of a family of yeomen, "free," as he says, "from the temptations of poverty and riches," [52] Baxter had naturally strong views as to the ethics of landowning. Significantly enough, he deals with them under the general rubric of "Cases of oppression, especially of tenants," oppression being defined as the "injuring of inferiors who are unable to resist or to right themselves." "It is too common a sort of oppression for the rich in all places to domineer too insolently over the poor, and force them to follow their wills and to serve their interest, be it right or wrong. . . . Especially unmerciful landlords are the common and sore oppressors of the countrymen. If a few men can but get money enough to purchase all the land in a county, they think that they may do with their own as they list, and set such hard bargains of it to their tenants, that they are all but as their servants. . . . An oppressor is an

Anti-Christ and an Anti-God . . . not only the agent of the Devil, but his image." As in his discussion of prices, the gist of Baxter's analysis of the cases of conscience which arise in the relations of landlord and tenant is that no man may secure pecuniary gain for himself by injuring his neighbor. Except in unusual circumstances, a landlord must not let his land at the full competitive rent which it would fetch in the market: "Ordinarily the common sort of tenants in England should have so much abated of the fullest worth that they may comfortably live on it, and follow their labours with cheerfulness of mind and liberty to serve God in their families, and to mind the matters of their salvation, and not to be necessitated to such toil and care and pinching want as shall make them liker slaves than free men." He must not improve (i.e., enclose) his land without considering the effect on the tenants, or evict his tenants without compensating them, and in such a way as to cause depopulation; nor must a newcomer take a holding over the sitting tenant's head by offering "a greater rent than he can give or than the landlord hath just cause to require of him." The Christian, in short, while eschewing "causeless, perplexing, melancholy scruples, which would stop a man in the course of his duty," must so manage his business as to "avoid sin rather than loss," and seek first to keep his conscience in peace.

The first characteristic to strike the modern reader in all this teaching is its conservatism. In spite of the economic and political revolutions of the past two centuries, how small, after all, the change in the presentation of the social ethics of the Christian faith! A few months after the appearance of the *Christian Directory*, the Stop of the Exchequer tore a hole in the already intricate web of London finance, and sent a shiver through the money-markets of Europe. But Baxter, though no mere antiquarian, discourses of equity in bargaining, of just prices, of reasonable rents, of the sin of usury, in the same tone, if not with quite the same conclusions, as a medieval Schoolman, and he differs from one of the later Doctors, like St. Antonino, hardly more than St. Antonino himself had differed from Aquinas.  Seven years later Bunyan published *The Life and Death of Mr. Badman.*  Among the vices which it pilloried were the sin of extortion, "most commonly committed by men of trade, who without all conscience, when they have an advantage, will make a prey of their neighbour," the covetousness of "hucksters, that buy up the poor man's victual wholesale and sell it to him again for unreasonable gains," the avarice of usurers, who watch till "the poor fall into their mouths," and "of those vile wretches called pawnbrokers, that lend money and goods to poor people, who are by necessity forced to such an inconvenience, and will make by one trick or another the interest of what they so lend amount to thirty and forty, yea, sometimes fifty pounds by the year."  As Christian and Christiana watched Mr. Badman thus bite and pinch the poor in his shop in Bedford, before they took staff and scrip for their journey to a more distant City, they remembered that the Lord himself will plead the cause of the afflicted against them that oppress them, and reflected, taught by the dealings of Ephron the son of Zohar, and of David with Ormon the Jebusite, that there is a "wickedness, as in selling too dear, so in buying too cheap." [53]  Brother Berthold of Regensburg had said the same four centuries before in his racy sermons in Germany. The emergence of the idea that "business is business," and that the world of commercial transactions is a closed compartment with laws of its own, if more ancient than is often supposed, did not win so painless a triumph as is sometimes suggested.  Puritan as well as Catholic accepted without demur the view which set all human interests and activities within the compass of religion.  Puritans, as well as Catholics, essayed the formidable task of formulating a Christian casuistry of economic conduct.

They essayed it. But they succeeded even less than the Popes and Doctors whose teaching, not always unwittingly, they repeated. And their failure had its roots, not merely in the obstacles offered by the ever more recalcitrant opposition of a commercial environment, but, like all failures which are significant, in the soul of Puritanism itself. Virtues are often conquered by vices, but their rout is most complete when it is inflicted by other virtues, more militant, more efficient, or more congenial, and it is not only tares which choke the ground where the good seed is sown. The fundamental question, after all, is not what kind of rules a faith enjoins, but what type of character it esteems and cultivates. To the scheme of Christian ethics which offered admonitions against the numberless disguises assumed by the sin which sticketh fast between buying and selling, the Puritan character offered, not direct opposition, but a polished surface on which these ghostly admonitions could find no enduring foothold. The rules of Christian morality elaborated by Baxter were subtle and sincere. But they were like seeds carried by birds from a distant and fertile plain, and dropped upon a glacier. They were at once embalmed and sterilized in a river of ice.

"The capitalist spirit" is as old as history, and was not, as has sometimes been said, the offspring of Puritanism. But it found in certain aspects of later Puritanism a tonic which braced its energies and fortified its already vigorous temper. At first sight, no contrast could be more violent than that between the iron collectivism, the almost military discipline, the remorseless and violent rigors practiced in Calvin's Geneva, and preached elsewhere, if in a milder form, by his disciples, and the impatient rejection of all traditional restrictions on economic enterprise which was the temper of the English business world after the Civil War.

In reality, the same ingredients were present throughout, but they were mixed in changing proportions, and exposed to different temperatures at different times.   Like traits of individual character which are suppressed till the approach of maturity releases them, the tendencies in Puritanism, which were to make it later a potent ally of the movement against the control of economic relations in the name either of social morality or of the public interest, did not reveal themselves till political and economic changes had prepared a congenial environment for their growth.   Nor, once those conditions were created, was it only England which witnessed the transformation.   In all countries alike, in Holland, in America, in Scotland, in Geneva itself, the social theory of Calvinism went through the same process of development.   It had begun by being the very soul of authoritarian regimentation. It ended by being the vehicle of an almost Utilitarian individualism.   While social reformers in the sixteenth century could praise Calvin for his economic rigor, their successors in Restoration England, if of one persuasion, denounced him as the parent of economic license, if of another, applauded Calvinist communities for their commercial enterprise and for their freedom from antiquated prejudices on the subject of economic morality.   So little do those who shoot the arrows of the spirit know where they will light.

### III.   THE TRIUMPH OF THE ECONOMIC VIRTUES

"One beam in a dark place," wrote one who knew the travail of the spirit, "hath exceeding much refreshment in it.   Blessed be His name for shining upon so dark a heart as mine." [54]   While the revelation of God to the individual soul is the center of all religion, the essence of Puritan theology was that it made it, not only the center, but the whole circumference and substance, dismissing as dross and vanity all else but this secret and solitary communion.   Grace alone can save, and this grace is the direct gift of God, unmediated by any earthly institution. The elect cannot by any act of their own evoke it; but they can prepare their hearts to receive it, and cherish it when received. They will prepare them best, if they empty them of all that may disturb the intentness of their lonely vigil. Like an engineer, who, to canalize the rush of the oncoming tide, dams all channels save that through which it is to pour, like a painter who makes light visible by plunging all that is not light in gloom, the Puritan attunes his heart to the voice from Heaven by an immense effort of concentration and abnegation. To win all, he renounces all. When earthly props have been cast down, the soul stands erect in the presence of God. Infinity is attained by a process of subtraction.

To a vision thus absorbed in a single intense experience, not only religious and ecclesiastical systems, but the entire world of human relations, the whole fabric of social institutions, witnessing in all the wealth of their idealism and their greed to the infinite creativeness of man, reveal themselves in a new and wintry light. The fire of the spirit burns brightly on the hearth; but through the windows of his soul the Puritan, unless a poet or a saint, looks on a landscape touched by no breath of spring. What he sees is a forbidding and frost-bound wilderness, rolling its snow-clad leagues towards the grave—a wilderness to be subdued with aching limbs beneath solitary stars. Through it he must take his way, alone. No aid can avail him: no preacher, for only the elect can apprehend with the spirit the word of God; no Church, for to the visible Church even reprobates belong; no sacrament, for sacraments are ordained to increase the glory of God, not to minister spiritual nourishment to man; hardly God himself, for Christ died for the elect, and it may well be that the majesty of the Creator is revealed by the eternal damnation of all but a remnant of the created.[55]

His life is that of a soldier in hostile territory. He suffers in spirit the perils which the first settlers in America endured in body, the sea behind, the untamed desert in front, a cloud of inhuman enemies on either hand. Where Catholic and Anglican had caught a glimpse of the invisible, hovering like a consecration over the gross world of sense, and touching its muddy vesture with the unearthly gleam of a divine, yet familiar, beauty, the Puritan mourned for a lost Paradise and a creation sunk in sin. Where they had seen society as a mystical body, compact of members varying in order and degree, but dignified by participation in the common life of Christendom, he saw a bleak antithesis between the spirit which quickeneth and an alien, indifferent or hostile world. Where they had reverenced the decent order whereby past was knit to present, and man to man, and man to God, through fellowship in works of charity, in festival and fast, in the prayers and ceremonies of the Church, he turned with horror from the filthy rags of human righteousness. Where they, in short, had found comfort in a sacrament, he started back from a snare set to entrap his soul.

> We receive but what we give,
> And in our life alone does Nature live.

Too often, contemning the external order as unspiritual, he made it, and ultimately himself, less spiritual by reason of his contempt.

Those who seek God in isolation from their fellowmen, unless trebly armed for the perils of the quest, are apt to find, not God, but a devil, whose countenance bears an embarrassing resemblance to their own. The moral self-sufficiency of the Puritan nerved his will, but it corroded his sense of social solidarity. For, if each individual's destiny hangs on a private transaction between himself and his Maker, what room is left for human intervention? A servant of Jehovah more than of Christ, he revered God as a Judge rather than loved him as a Father, and was moved less by compassion for his erring brethren than by impatient indignation at the blindness of vessels of wrath who "sinned their mercies." A spiritual aristocrat, who sacrificed fraternity to liberty, he drew from his idealization of personal responsibility a theory of individual rights, which, secularized and generalized, was to be among the most potent explosives that the world has known. He drew from it also a scale of ethical values, in which the traditional scheme of Christian virtues was almost exactly reversed, and which, since he was above all things practical, he carried as a dynamic into the routine of business and political life.

For, since conduct and action, though availing nothing to attain the free gift of salvation, are a proof that the gift has been accorded, what is rejected as a means is resumed as a consequence, and the Puritan flings himself into practical activities with the dæmonic energy of one who, all doubts allayed, is conscious that he is a sealed and chosen vessel. Once engaged in affairs, he brings to them both the qualities and limitations of his creed in all their remorseless logic. Called by God to labor in his vineyard, he has within himself a principle at once of energy and of order, which makes him irresistible both in war and in the struggles of commerce. Convinced that character is all and circumstances nothing, he sees in the poverty of those who fall by the way, not a misfortune to be pitied and relieved, but a moral failing to be condemned, and in riches, not an object of suspicion—though like other gifts they may be abused—but the blessing which rewards the triumph of energy and will. Tempered by self-examination, self-discipline, self-control, he is the practical ascetic, whose victories are won not in the cloister, but on the battlefield, in the counting-house, and in the market.

This temper, of course with infinite varieties of quality

and emphasis, found its social organ in those middle and commercial classes who were the citadel of the Puritan spirit, and whom, "ennobled by their own industry and virtue," [56] Milton described as the standard-bearers of progress and enlightenment.  We are so accustomed to think of England as *par excellence* the pioneer of economic progress, that we are apt to forget how recently that rôle has been assumed.  In the Middle Ages it belonged to the Italians, in the sixteenth century to the Netherland dominions of the Spanish Empire, in the seventeenth to the United Provinces and, above all, to the Dutch.

The England of Shakespeare and Bacon was still largely medieval in its economic organization and social outlook, more interested in maintaining customary standards of consumption than in accumulating capital for future production, with an aristocracy contemptuous of the economic virtues, a peasantry farming for subsistence amid the organized confusion of the open-field village, and a small, if growing, body of jealously conservative craftsmen.  In such a society Puritanism worked like the yeast which sets the whole mass fermenting.  It went through its slack and loosely knit texture like a troop of Cromwell's Ironsides through the disorderly cavalry of Rupert.  Where, as in Ireland, the elements were so alien that assimilation was out of the question, the result was a wound that festered for three centuries.  In England the effect was that at once of an irritant and of a tonic.  Puritanism had its own standards of social conduct, derived partly from the obvious interests of the commercial classes, partly from its conception of the nature of God and the destiny of man.  These standards were in sharp antithesis, both to the considerable surviving elements of feudalism in English society, and to the policy of the authoritarian State, with its ideal of an ordered and graded society, whose different members were to be maintained in their traditional status by the pressure and protection of a paternal monarchy.  Sapping the former by its influence and overthrowing the latter by direct attack, Puritanism became a potent force in preparing the way for the commercial civilization which finally triumphed at the Revolution.

The complaint that religious radicalism, which aimed at upsetting the government of the Church, went hand in hand with an economic radicalism, which resented the restraints on individual self-interest imposed in the name of religion or of social policy, was being made by the stricter school of religious opinion quite early in the reign of Elizabeth.[57] Seventeenth-century writers repeated the charge that the Puritan conscience lost its delicacy where matters of business were concerned, and some of them were sufficiently struck by the phenomenon to attempt an historical explanation of it.  The example on which they usually seized—the symbol of a supposed general disposition to laxity—was the indulgence shown by Puritan divines in the particular matter of moderate interest.  It was the effect, so the picturesque story ran,[58] of the Marian persecution.  The refugees who fled to the Continent could not start business in a foreign country.  If, driven by necessity, they invested their capital and lived on the proceeds, who could quarrel with so venial a lapse in so good a cause?  Subsequent writers embellished the picture.  The redistribution of property at the time of the Dissolution, and the expansion of trade in the middle of the century, had led, one of them argued, to a great increase in the volume of credit transactions. The opprobrium which attached to loans at interest—"a sly and forbid practice"—not only among Romanists and Anglicans, but among honest Puritans, played into the hands of the less scrupulous members of "the faction."  Disappointed in politics, they took to money-lending, and, without venturing to justify usury in theory, defended it in practice.  "Without the scandal of a recantation, they contrived an expedient, by maintaining that, though usury for the name were stark naught, yet for widows, orphans and other impotents (therein principally comprising the saints under persecution) it was very tolerable, because profitable, and in a manner necessary." Naturally, Calvin's doctrine as to the legitimacy of moderate interest was hailed by these hypocrites with a shout of glee. "It took with the brethren like polygamy with the Turks, recommended by the example of divers zealous ministers, who themselves desired to pass for orphans of the first rank." [59]  Nor was it only as the apologist of moderate interest that Puritanism was alleged to reveal the cloven hoof. Puritans themselves complained of a mercilessness in driving hard bargains, and of a harshness to the poor, which contrasted unfavorably with the practice of followers of the unreformed religion. "The Papists," wrote a Puritan in 1653, "may rise up against many of this generation. It is a sad thing that they should be more forward upon a bad principle than a Christian upon a good one." [60]

Such, in all ages, is history as seen by the political pamphleteer. The real story was less dramatic, but more significant. From the very beginning, Calvinism had comprised two elements, which Calvin himself had fused, but which contained the seeds of future discord. It had at once given a whole-hearted *imprimatur* to the life of business enterprise, which most earlier moralists had regarded with suspicion, and had laid upon it the restraining hand of an inquisitorial discipline. At Geneva, where Calvinism was the creed of a small and homogeneous city, the second aspect had predominated; in the many-sided life of England, where there were numerous conflicting interests to balance it, and where it was long politically weak, the first. Then, in the late sixteenth and early seventeenth centuries, had come the wave of commercial and financial expansion—companies, colonies, capitalism in textiles, capitalism in mining, capitalism in finance—on the crest of which the English commercial classes, in Calvin's day still held in leading-strings by conservative statesmen, had climbed to a position of dignity and affluence.

Naturally, as the Puritan movement came to its own, these two elements flew apart. The collectivist, half-communistic aspect, which had never been acclimatized in England, quietly dropped out of notice, to crop up once more, and for the last time, to the disgust and terror of merchant and landowner, in the popular agitation under the Commonwealth. The individualism congenial to the world of business became the distinctive characteristic of a Puritanism which had arrived, and which, in becoming a political force, was at once secularized and committed to a career of compromise. Its note was not the attempt to establish on earth a "Kingdom of Christ," but an ideal of personal character and conduct, to be realized by the punctual discharge both of public and private duties. Its theory had been discipline; its practical result was liberty.

Given the social and political conditions of England, the transformation was inevitable. The incompatibility of Presbyterianism with the stratified arrangement of English society had been remarked by Hooker.[61] If the City Fathers of Geneva had thrown off by the beginning of the seventeenth century the religious collectivism of Calvin's régime, it was not to be expected that the landowners and *bourgeoisie* of an aristocratic and increasingly commercial nation, however much Calvinist theology might appeal to them, would view with favor the social doctrines implied in Calvinist discipline. In the reign of the first two Stuarts, both economic interests and political theory pulled them hard in the opposite direction. "Merchants' doings," the man of business in Wilson's *Discourse upon Usury* had observed, "must not thus be overthwarted by preachers and others that cannot skill of their dealings." [62]    Behind the elaborate façade of Tudor State control, which has attracted the attention of historians, an individualist movement had been steadily developing, which found expression in opposition to the traditional policy of stereotyping economic relations by checking enclosure, controlling food supplies and prices, interfering with the money-market, and regulating the conditions of the wage contract and of apprenticeship. In the first forty years of the seventeenth century, on grounds both of expediency and of principle, the commercial and propertied classes were becoming increasingly restive under the whole system, at once ambitious and inefficient, of economic paternalism.   It was in the same sections of the community that both religious and economic dissatisfaction were most acute.   Puritanism, with its idealization of the spiritual energies which found expression in the activities of business and industry, drew the isolated rivulets of discontent together, and swept them forward with the dignity and momentum of a religious and a social philosophy.

For it was not merely as the exponent of certain tenets as to theology and church government, but as the champion of interests and opinions embracing every side of the life of society, that the Puritan movement came into collision with the Crown.   In reality, as is the case with most heroic ideologies, the social and religious aspects of Puritanism were not disentangled; they presented themselves, both to supporters and opponents, as different facets of a single scheme. "All that crossed the views of the needy courtiers, the proud encroaching priests, the thievish projectors, the lewd nobility and gentry . . . whoever could endure a sermon, modest habit or conversation, or anything good—all these were Puritans." [63]   The clash was not one of theories —a systematic and theoretical individualism did not develop till after the Restoration—but of contradictory economic interests and incompatible conceptions of social expediency.

The economic policy haltingly pursued by the Government

of Charles I bore some resemblance to the system of which a more uncompromising version was developed between 1661 and 1685 by Colbert in France. It was one which favored an artificial and State-promoted capitalism—a capitalism resting on the grant of privileges and concessions to company promoters who would pay for them, and accompanied by an elaborate system of State control, which again, if partly inspired by a genuine solicitude for the public interest, was too often smeared with an odious trail of finance. It found its characteristic expression in the grant of patents, in the revival of the royal monopoly of exchange business, against which the City had fought under Elizabeth, in attempts to enforce by administrative action compliance with the elaborate and impracticable code controlling the textile trades and to put down speculation in foodstuffs, and in raids on enclosing landlords, on employers who paid in truck or evaded the rates fixed by assessment, and on justices who were negligent in the administration of the Poor Laws. Such measures were combined with occasional plunges into even more grandiose schemes for the establishment of county granaries, for taking certain industries into the hands of the Crown, and even for the virtual nationalization of the cloth manufacture. [64]

"The very genius of that nation of people," wrote Strafford to Laud of the Puritans, "leads them always to oppose, as well civilly as ecclesiastically, all that ever authority ordains for them." [65] Against this whole attempt to convert economic activity into an instrument of profit for the Government and its hangers-on—against, no less, the spasmodic attempts of the State to protect peasants against landlords, craftsmen against merchants, and consumers against middlemen—the interests which it thwarted and curbed revolted with increasing pertinacity. Questions of taxation, on which attention has usually been concentrated, were in reality merely one element in a quarrel which had its deeper cause in the collision of incompatible social philosophies. The Puritan tradesman had seen his business ruined by a monopoly granted to a needy courtier, and cursed Laud and his Popish soap. The Puritan goldsmith or financier had found his trade as a bullion-broker hampered by the reëstablishment of the ancient office of Royal Exchanger, and secured a resolution from the House of Commons, declaring that the patent vesting it in Lord Holland and the proclamation forbidding the exchanging of gold and silver by unauthorized persons were a grievance. The Puritan money-lender had been punished by the Court of High Commission, and railed at the interference of bishops in temporal affairs. The Puritan clothier, who had suffered many things at the hands of interfering busy-bodies despatched from Whitehall to teach him his business, averted discreet eyes when the Wiltshire workmen threw a more than usually obnoxious Royal Commissioner into the Avon, and, when the Civil War came, rallied to the Parliament. The Puritan country gentleman had been harried by Depopulation Commissions, and took his revenge with the meeting of the Long Parliament. The Puritan merchant had seen the Crown both squeeze money out of his company, and threaten its monopoly by encouraging courtly interlopers to infringe its charter. The Puritan member of Parliament had invested in colonial enterprises, and had ideas as to commercial policy which were not those of the Government. Confident in their own energy and acumen, proud of their success, and regarding with profound distrust the interference both of Church and of State with matters of business and property rights, the commercial classes, in spite of their attachment to a militant mercantilism in matters of trade, were, even before the Civil War, more than half converted to the administrative nihilism which was to be the rule of social policy in the century following it. Their demand was the one which is usual in such circumstances.

It was that business affairs should be left to be settled by business men, unhampered by the intrusions of an antiquated morality or by misconceived arguments of public policy.[66]

The separation of economic from ethical interests, which was the note of all this movement, was in sharp opposition to religious tradition, and it did not establish itself without a struggle. Even in the very capital of European commerce and finance, an embittered controversy was occasioned by the refusal to admit usurers to communion or to confer degrees upon them; it was only after a storm of pamphleteering, in which the theological faculty of the University of Utrecht performed prodigies of zeal and ingenuity, that the States of Holland and West Friesland closed the agitation by declaring that the Church had no concern with questions of banking.[67] In the French Calvinist Churches, the decline of discipline had caused lamentations a generation earlier.[68] In America, the theocracy of Massachusetts, merciless alike to religious liberty and to economic license, was about to be undermined by the rise of new States like Rhode Island and Pennsylvania, whose tolerant, individualist and utilitarian temper was destined to find its greatest representative in the golden common sense of Benjamin Franklin.[69] "The sin of our too great fondness for trade, to the neglecting of our more valuable interests," wrote a Scottish divine in 1709, when Glasgow was on the eve of a triumphant outburst of commercial enterprise, "I humbly think will be written upon our judgment. . . . I am sure the Lord is remarkably frowning upon our trade . . . since it was put in the room of religion." [70]

In England, the growing disposition to apply exclusively economic standards to social relations evoked from Puritan writers and divines vigorous protests against usurious interest, extortionate prices and the oppression of tenants by landlords. The faithful, it was urged, had interpreted only too literally the doctrine that the sinner was saved, not by works, but by faith. Usury, "in time of Popery an odious thing," [71] had become a scandal. Professors, by their covetousness, caused the enemies of the reformed religion to blaspheme.[72] The exactions of the forestaller and regrater were never so monstrous or so immune from interference. The hearts of the rich were never so hard, nor the necessities of the poor so neglected. "The poor able to work are suffered to beg; the impotent, aged and sick are not sufficiently provided for, but almost starved with the allowance of 3d. and 4d. a piece a week. . . . These are the last times indeed. Men generally are all for themselves. And some would set up such, having a form of religion, without the power of it." [73]

These utterances came, however, from that part of the Puritan mind which looked backward. That which looked forward found in the rapidly growing spirit of economic enterprise something not uncongenial to its own temper, and went out to welcome it as an ally. What in Calvin had been a qualified concession to practical exigencies appeared in some of his later followers as a frank idealization of the life of the trader, as the service of God and the training-ground of the soul. Discarding the suspicion of economic motives, which had been as characteristic of the reformers as of medieval theologians, Puritanism in its later phases added a halo of ethical sanctification to the appeal of economic expediency, and offered a moral creed, in which the duties of religion and the calls of business ended their long estrangement in an unanticipated reconciliation. Its spokesmen pointed out, it is true, the peril to the soul involved in a single-minded concentration on economic interests. The enemy, however, was not riches, but the bad habits sometimes associated with them, and its warnings against an excessive preoccupation with the pursuit of gain wore more and more the air of after-thoughts, appended to teaching the main tendency and emphasis of which were little affected by these incidental qualifications. It insisted, in short, that money-making, if not free from spiritual dangers, was not a danger and nothing else, but that it could be, and ought to be, carried on for the greater glory of God.

The conception to which it appealed to bridge the gulf sprang from the very heart of Puritan theology. It was that expressed in the characteristic and oft-used phrase, "a Calling." [74] The rational order of the universe is the work of God, and its plan requires that the individual should labor for God's glory. There is a spiritual calling, and a temporal calling. It is the first duty of the Christian to know and believe in God; it is by faith that he will be saved. But faith is not a mere profession, such as that of Talkative of Prating Row, whose "religion is to make a noise." The only genuine faith is the faith which produces works. "At the day of Doom men shall be judged according to their fruits. It will not be said then, Did you believe? but, Were you doers, or talkers only?" [75] The second duty of the Christian is to labor in the affairs of practical life, and this second duty is subordinate only to the first. "God," wrote a Puritan divine, "doth call every man and woman . . . to serve him in some peculiar employment in this world, both for their own and the common good. . . . The Great Governour of the world hath appointed to every man his proper post and province, and let him be never so active out of his sphere, he will be at a great loss, if he do not keep his own vineyard and mind his own business." [76]

From this reiterated insistence on secular obligations as imposed by the divine will, it follows that, not withdrawal from the world, but the conscientious discharge of the duties of business, is among the loftiest of religious and moral virtues. "The begging friars and such monks as live only to themselves and to their formal devotion, but do employ themselves in no one thing to further their own subsistence or the good of mankind . . . yet have the confidence to boast of this their course as a state of perfection; which in very deed, as to the worthiness of it, falls short of the poorest cobbler, for his is a calling of God, and theirs is none." [77] The idea was not a new one. Luther had advanced it as a weapon against monasticism. But for Luther, with his patriarchal outlook on economic affairs, the calling means normally that state of life in which the individual has been set by Heaven, and against which it is impiety to rebel. On the lips of Puritan divines, it is not an invitation to resignation, but the bugle-call which summons the elect to the long battle which will end only with their death. "The world is all before them." They are to hammer out their salvation, not merely *in vocatione,* but *per vocationem.* The calling is not a condition in which the individual is born, but a strenuous and exacting enterprise, to be undertaken, indeed, under the guidance of Providence, but to be chosen by each man for himself, with a deep sense of his solemn responsibilities. "God hath given to man reason for this use, that he should first consider, then choose, then put in execution; and it is a preposterous and brutish thing to fix or fall upon any weighty business, such as a calling or condition of life, without a careful pondering it in the balance of sound reason." [78]

*Laborare est orare.* By the Puritan moralist the ancient maxim is repeated with a new and intenser significance. The labor which he idealizes is not simply a requirement imposed by nature, or a punishment for the sin of Adam. It is itself a kind of ascetic discipline, more rigorous than that demanded of any order of mendicants—a discipline imposed by the will of God, and to be undergone, not in solitude, but in the punctual discharge of secular duties. It is not merely an economic means, to be laid aside when physical needs have been satisfied. It is a spiritual end, for in it alone can the soul find health, and it must be continued as an ethical duty long after it has ceased to be a material necessity. Work thus conceived stands at the very opposite pole from "good works," as they were understood, or misunderstood, by Protestants. They, it was thought, had been a series of single transactions, performed as compensation for particular sins, or out of anxiety to acquire merit. What is required of the Puritan is not individual meritorious acts, but a holy life—a system in which every element is grouped round a central idea, the service of God, from which all disturbing irrelevances have been pruned, and to which all minor interests are subordinated.

His conception of that life was expressed in the words, "Be wholly taken up in diligent business of your lawful callings, when you are not exercised in the more immediate service of God." [79] In order to deepen his spiritual life, the Christian must be prepared to narrow it. He "is blind in no man's cause, but best sighted in his own. He confines himself to the circle of his own affairs and thrusts not his fingers in needless fires. . . . He sees the falseness of it [the world] and therefore learns to trust himself ever, others so far as not to be damaged by their disappointment." [80] There must be no idle leisure: "those that are prodigal of their time despise their own souls." [81] Religion must be active, not merely contemplative. Contemplation is, indeed, a kind of self-indulgence. "To neglect this [i.e., bodily employment and mental labor] and say, 'I will pray and meditate,' is as if your servant should refuse your greatest work, and tye himself to some lesser, easie part. . . . God hath commanded you some way or other to labour for your daily bread." [82] The rich are no more excused from work than the poor, though they may rightly use their riches to select some occupation specially serviceable to others. Covetousness is a danger to the soul, but it is not so grave a danger as sloth. "The standing pool is prone to putrefaction: and it were better to beat down the body and to keep it in subjection by a laborious calling, than through luxury to become a cast-away." [83]   So far from poverty being meritorious, it is a duty to choose the more profitable occupation.  "If God show you a way in which you may lawfully get more than in another way (without wrong to your soul or to any other), if you refuse this, and choose the less gainful way, you cross one of the ends of your Calling, and you refuse to be God's steward."   Luxury, unrestrained pleasure, personal extravagance, can have no place in a Christian's conduct, for "every penny which is laid out . . . must be done as by God's own appointment."   Even excessive devotion to friends and relations is to be avoided.   "It is an irrational act, and therefore not fit for a rational creature, to love any one farther than reason will allow us. . . . It very often taketh up men's minds so as to hinder their love to God." [84]   The Christian life, in short, must be systematic and organized, the work of an iron will and a cool intelligence.   Those who have read Mill's account of his father must have been struck by the extent to which Utilitarianism was not merely a political doctrine, but a moral attitude. Some of the links in the Utilitarian coat of mail were forged, it may be suggested, by the Puritan divines of the seventeenth century.

The practical application of these generalities to business is set out in the numerous works composed to expound the rules of Christian conduct in the varied relations of life.   If one may judge by their titles—*Navigation Spiritualized, Husbandry Spiritualized, The Religious Weaver* [85]—there must have been a considerable demand for books conducive to professional edification.   A characteristic specimen is *The Tradesman's Calling*,[86] by Richard Steele.   The author, after being deprived of a country living under the Act of Uniformity, spent his declining years as minister of a congregation at Armourers Hall in London, and may be presumed to have understood the spiritual requirements of the City in his day, when the heroic age of Puritanism was almost over and enthusiasm was no longer a virtue. No one who was writing a treatise on economic ethics today would address himself primarily to the independent shopkeeper, as the figure most representative of the business community, and Steele's book throws a flood of light on the problems and outlook of the *bourgeoisie*, in an age before the center of economic gravity had shifted from the substantial tradesman to the exporting merchant, the industrial capitalist and the financier.

Like Baxter, he is acquainted with the teaching of earlier authorities as to equity in bargaining. He is doubtful, however, of its practical utility. Obvious frauds in matters of quality and weight are to be avoided; an honest tradesman ought not to corner the market, or "accumulate two or three callings merely to increase his riches," or oppress the poor; nor should he seek more than "a reasonable proportion of gain," or "lie on the catch to make [his] markets of others' straits." But Steele rejects as useless in practice the various objective standards of a reasonable profit—cost of production, standard of life, customary prices—which had been suggested in earlier ages, and concludes that the individual must judge for himself. "Here, as in many other cases, an upright conscience must be the clerk of the market."

In reality, however, the characteristic of *The Tradesman's Calling*, as of the age in which it was written, is not the relics of medieval doctrine which linger embalmed in its guileless pages, but the robust common sense, which carries the author lightly over traditional scruples on a tide of genial, if Philistine, optimism. For his main thesis is a comfortable one—that there is no necessary conflict between religion and business. "Prudence and Piety were always very good friends. . . . You may gain enough of both worlds if you would mind each in its place." His object is to show how that agreeable result may be produced by dedicating business—with due reservations—to the service of God, and he has naturally little to say on the moral casuistry of economic conduct, because he is permeated by the idea that trade itself is a kind of religion. A tradesman's first duty is to get a full insight into his calling, and to use his brains to improve it. "He that hath lent you talents hath also said, 'Occupy till I come!' Your strength is a talent, your parts are talents, and so is your time. How is it that ye stand all the day idle? . . . Your trade is your proper province. . . . Your own vineyard you should keep. . . . Your fancies, your understandings, your memories . . . are all to be laid out therein." So far from their being an inevitable collision between the requirements of business and the claims of religion, they walk hand in hand. By a fortunate dispensation, the virtues enjoined on Christians—diligence, moderation, sobriety, thrift—are the very qualities most conducive to commercial success. The foundation of all is prudence; and prudence is merely another name for the "godly wisdom [which] comes in and puts due bounds" to his expenses, "and teaches the tradesman to live rather somewhat below than at all above his income." Industry comes next, and industry is at once expedient and meritorious. It will keep the tradesman from "frequent and needless frequenting of taverns," and pin him to his shop, "where you may most confidently expect the presence and blessing of God."

If virtue is advantageous, vice is ruinous. Bad company, speculation, gambling, politics, and "a preposterous zeal" in religion—it is these things which are the ruin of tradesmen. Not, indeed, that religion is to be neglected. On the contrary, it "is to be exercised in the frequent use of holy ejaculations." What is deprecated is merely the unbusiness-like habit of "neglecting a man's necessary affairs upon pretence of religious worship." But these faults, common and uncommon alike, are precisely those to be avoided by the sincere Christian, who must not, indeed, deceive or oppress his neighbor, but need not fly to the other extreme, be righteous overmuch, or refuse to "take the advantage which the Providence of God puts into his hands." By a kind of happy, preëstablished harmony, such as a later age discovered between the needs of society and the self-interest of the individual, success in business is in itself almost a sign of spiritual grace, for it is a proof that a man has labored faithfully in his vocation, and that "God has blessed his trade." "Nothing will pass in any man's account except it be done in the way of his calling. . . . Next to the saving his soul, [the tradesman's] care and business is to serve God in his calling, and to drive it as far as it will go."

When duty was so profitable, might not profit-making be a duty? Thus argued the honest pupils of Mr. Gripeman, the schoolmaster of Love-gain, a market-town in the county of Coveting in the north.[87] The inference was illogical, but how attractive! When the Rev. David Jones was so indiscreet as to preach at St. Mary Woolnoth in Lombard Street a sermon against usury on the text, "The Pharisees who were covetous heard all these things and they derided Christ," his career in London was brought to an abrupt conclusion.[88]

The springs of economic conduct lie in regions rarely penetrated by moralists, and to suggest a direct reaction of theory on practice would be paradoxical. But, if the circumstances which determine that certain kinds of conduct shall be profitable are economic, those which decide that they shall be the object of general approval are primarily moral and intellectual. For conventions to be adopted with wholehearted enthusiasm, to be not merely tolerated, but applauded, to become the habit of a nation and the admiration of its philosophers, the second condition must be present as well as the first. The insistence among men of pecuniary motives, the strength of economic egotism, the appetite for gain—these are the commonplaces of every age and need no emphasis. What is significant is the change of standards which converted a natural frailty into a resounding virtue. After all, it appears, a man can serve two masters, for—so happily is the world disposed—he may be paid by one, while he works for the other. Between the old-fashioned denunciation of uncharitable covetousness and the new-fashioned applause of economic enterprise, a bridge is thrown by the argument which urges that enterprise itself is the discharge of a duty imposed by God.

In the year 1690 appeared a pamphlet entitled *A Discourse of Trade, by N. B., M.D.*[89] Notable for its enlightened discussion of conventional theories of the balance of trade, it is a good specimen of an indifferent *genus.* But its authorship was more significant than its argument. For N. B. was Dr. Nicholas Barbon; and Dr. Nicholas Barbon, currency expert, pioneer of insurance, and enthusiast for landbanks, was the son of that Praise-God Barebones, by the parody of whose alluring surname a cynical posterity recorded its verdict on the brief comedy of the Rule of the Saints over Laodicean Englishmen. The reaction from Puritan rigor to Restoration license is the most familiar of platitudes. The reaction to a mundane materialism was more gradual, more general, and ultimately of greater significance. The profligacy of the courtier had its decorous counterpart in the economic orgies of the tradesman and the merchant. Votaries, not of Bacchus, but of a more exacting and more profitable divinity, they celebrated their relief at the discredit of a too arduous idealism, by plunging with redoubled zest into the agreeable fever of making and losing money.

The transition from the anabaptist to the company promoter was less abrupt than might at first sight be supposed. It had been prepared, however unintentionally, by Puritan moralists. In their emphasis on the moral duty of untiring

activity, on work as an end in itself, on the evils of luxury
and extravagance, on foresight and thrift, on moderation
and self-discipline and rational calculation, they had created
an ideal of Christian conduct, which canonized as an ethical
principle the efficiency which economic theorists were
preaching as a specific for social disorders. It was as capti-
vating as it was novel. To countless generations of religious thinkers, the fundamental maxim of Christian social
ethics had seemed to be expressed in the words of St. Paul to
Timothy: "Having food and raiment, let us be therewith
content. For the love of money is the root of all evil."
Now, while, as always, the world battered at the gate, a
new standard was raised within the citadel by its own defenders. The garrison had discovered that the invading
host of economic appetites was, not an enemy, but an ally.
Not sufficiency to the needs of daily life, but limitless increase and expansion, became the goal of the Christian's efforts. Not consumption, on which the eyes of earlier sages
had been turned, but production, became the pivot of his
argument. Not an easy-going and open-handed charity,
but a systematic and methodical accumulation, won the meed
of praise that belongs to the good and faithful servant.
The shrewd, calculating commercialism which tries all human relations by pecuniary standards, the acquisitiveness
which cannot rest while there are competitors to be conquered or profits to be won, the love of social power and
hunger for economic gain—these irrepressible appetites had
evoked from time immemorial the warnings and denunciations of saints and sages. Plunged in the cleansing waters
of later Puritanism, the qualities which less enlightened ages
had denounced as social vices emerged as economic virtues.
They emerged as moral virtues as well. For the world exists not to be enjoyed, but to be conquered. Only its conqueror deserves the name of Christian. For such a philosophy, the question, "What shall it profit a man?" carries

no sting. In winning the world, he wins the salvation of his own soul as well.

The idea of economic progress as an end to be consciously sought, while ever receding, had been unfamiliar to most earlier generations of Englishmen, in which the theme of moralists had been the danger of unbridled cupidity, and the main aim of public policy had been the stability of traditional relationships. It found a new sanction in the identification of labor and enterprise with the service of God. The magnificent energy which changed in a century the face of material civilization was to draw nourishment from that temper. The worship of production and ever greater production—the slavish drudgery of the millionaire and his unhappy servants—was to be hallowed by the precepts of the same compelling creed.

Social development moves with a logic whose inferences are long delayed, and the day of these remoter applications had not yet dawned. The version of Christian ethics expounded by Puritanism in some of its later phases was still only in its vigorous youth. But it sailed forward on a flowing tide. It had an unconscious ally in the pre-occupation with economic interests which found expression in the enthusiasm of business politicians for a commercial *Machtpolitik*. The youthful Commonwealth, a rival of Holland "for the fairest mistress in the world—trade," [90] was not two years old when it made its own essay in economic imperialism. "A bare-faced war" for commerce, got up by the Royal African Company, was Clarendon's verdict [91] on the Dutch war of 1665-7. Five years later, Shaftesbury hounded the City against Holland with the cry of *Delenda est Carthago*. The war finance of the Protectorate had made it necessary for Cromwell to court Dutch and Jewish, as well as native, capitalists, and the impecunious Government of the Restoration was in the hands of those syndicates of goldsmiths whose rapacity the Chancellor, a survivor from the age before the deluge, when aristocrats still despised the upstart plutocracy, found not a little disgusting.[92]

The contemporary progress of economic thought fortified no less the mood which glorified the economic virtues. Economic science developed in England, not, as in Germany, as the handmaid of public administration, nor, as in France, through the speculations of philosophers and men of letters, but as the interpreter of the practical interests of the City. With the exception of Petty and Locke, its most eminent practitioners were business men, and the questions which excited them were those, neither of production nor of social organization, but of commerce and finance—the balance of trade, tariffs, interest, currency and credit. The rise of Political Arithmetic after the Restoration, profoundly influenced, as it was, by the Cartesian philosophy and by the progress of natural science, stamped their spontaneous and doctrineless individualism with the seal of theoretical orthodoxy. "Knowledge," wrote the author of the preface to a work by one of the most eminent exponents of the new science, "in great measure is become mechanical." [93] The exact analysis of natural conditions, the calculations of forces and strains, the reduction of the complex to the operation of simple, constant and measurable forces, was the natural bias of an age interested primarily in mathematics and physics. Its object was "to express itself in terms of number, weight or measure, to use only arguments of sense, and to consider only such causes as have visible foundations in nature; leaving those that depend upon the mutable minds, opinions, appetites and passions of particular men to the consideration of others." [94]

In such an atmosphere, the moral casuistry, which had occupied so large a place in the earlier treatment of social and economic subjects, seemed the voice of an antiquated superstition. Moreover, the main economic dogma of the mercantilist had an affinity with the main ethical dogma of the Puritan, which was the more striking because the coincidence was undesigned. To the former, production, not consumption, was the pivot of the economic system, and, by what seems to the modern reader a curious perversion, consumption is applauded only because it offers a new market for productive energies. To the latter, the cardinal virtues are precisely those which find in the strenuous toils of industry and commerce their most natural expression. The typical qualities of the successful business life, in the days before the rise of joint-stock enterprise, were intensity and earnestness of labor, concentration, system and method, the initiative which broke with routine and the foresight which postponed the present to the future. Advice like that of the Reverend Mr. Steele to his City congregation was admirably calculated to give these arduous excellences a heightened status and justification. The lean goddess, Abstinence. whom Mr. Keynes, in a passage of brilliant indiscretion, has revealed as the tutelary divinity of Victorian England, was inducted to the austere splendors of her ascetic shrine by the pious hands of Puritan moralists.

Such teaching fell upon willing ears. Excluded by legislation from a direct participation in public affairs, Dissenters of means and social position threw themselves into the alternative career offered by commerce and finance, and did so the more readily because religion itself had blessed their choice. If they conformed, the character given them by their critics—"opinionating, relying much upon their own judgment . . . ungrateful, as not holding themselves beholden to any man . . . proud, as thinking themselves the only favorites of God, and the only wise or virtuous among men" [95]—disposed them to the left in questions of Church and State. The names of the commercial magnates of the day lend some confirmation to the suggestion of that affinity between religious radicalism and business acumen which envious contemporaries expressed in their sneers at the "Presbyterian old usurer," "devout misers," and "extorting Ishban."[96]   The four London members elected in 1661 had not only filled the ordinary civic offices, but had held between them the governorship of the East India Company, the deputy-governorship of the Levant Company, and the masterships of the Salters and Drapers Companies; two of them were said to be Presbyterians, and two Independents.[97] Of the committee of leading business men who advised Charles II's Government on questions of commercial policy, some, like Sir Patience Ward and Michael Godfrey, represented the ultra-Protestantism of the City, while others, like Thomas Papillon and the two Houblons, were members of the French Huguenot church in London.[98]   In spite of the bitter commercial rivalry with Holland, both Dutch capital and Dutch ideas found an enthusiastic welcome in London.[99]   Sir George Downing, Charles II's envoy at the Hague, who endeavored to acclimatize Dutch banking methods in England, and who, according to Clarendon, was one of the intriguers who prepared the war of 1665-7, had been reared in the Puritan severity of Salem and Harvard, and had been a preacher in the regiment of Colonel Okey.[100] Paterson, who supplied the idea of a joint-stock banking corporation, which Michael Godfrey popularized in the City and Montagu piloted through Parliament, was, like the magnificent Law, a Scotch company promoter, who had haunted the Hague in the days when it was the home of disconsolate Whigs.[101]   Yarranton, most ingenious of projectors, had been an officer in the Parliamentary army, and his book was a long sermon on the virtues of the Dutch.[102]   Defoe, who wrote the idyll of the *bourgeoisie* in his *Complete English Tradesman,* was born of nonconformist parents, and was intended for the ministry before, having failed in trade, he took up politics and literature.[102]   In his admirable study of the iron industry, Mr. Ashton has shown that the most eminent iron-masters of the eighteenth century belonged as a rule to the Puritan connection.[104]   They had their prototype in the seventeenth century in Baxter's friend, Thomas Foley, "who from almost nothing did get about £5,000 per annum or more by iron works." [105]

To such a generation, a creed which transformed the acquisition of wealth from a drudgery or a temptation into a moral duty was the milk of lions.   It was not that religion was expelled from practical life, but that religion itself gave it a foundation of granite.   In that keen atmosphere of economic enterprise, the ethics of the Puritan bore some resemblance to those associated later with the name of Smiles. The good Christian was not wholly dissimilar from the economic man.

## IV.   THE NEW MEDICINE FOR POVERTY

To applaud certain qualities is by implication to condemn the habits and institutions which appear to conflict with them.   The recognition accorded by Puritan ethics to the economic virtues, in an age when such virtues were rarer than they are today, gave a timely stimulus to economic efficiency.   But it naturally, if unintentionally, modified the traditional attitude towards social obligations.   For the spontaneous, doctrineless individualism, which became the rule of English public life a century before the philosophy of it was propounded by Adam Smith, no single cause was responsible.   But, simultaneously with the obvious movements in the world of affairs—the discrediting of the ideal of a paternal, authoritarian Government, the breakdown of central control over local administration, the dislocation caused by the Civil War, the expansion of trade and the shifting of industry from its accustomed seats—it is perhaps not fanciful to detect in the ethics of Puritanism one force contributing to the change in social policy which is noticeable after the middle of the century.

The loftiest teaching cannot escape from its own shadow. To urge that the Christian life must be lived in a zealous discharge of private duties—how necessary! Yet how readily perverted to the suggestion that there are no vital social obligations beyond and above them! To insist that the individual is responsible, that no man can save his brother, that the essence of religion is the contact of the soul with its Maker, how true and indispensable! But how easy to slip from that truth into the suggestion that society is without responsibility, that no man can help his brother, that the social order and its consequences are not even the scaffolding by which men may climb to greater heights, but something external, alien and irrelevant—something, at best, indifferent to the life of the spirit, and, at worse, the sphere of the letter which killeth and of the reliance on works which ensnares the soul into the slumber of death! In emphasizing that God's Kingdom is not of this world, Puritanism did not always escape the suggestion that this world is no part of God's Kingdom. The complacent victim of that false antithesis between the social mechanism and the life of the spirit, which was to tyrannize over English religious thought for the next two centuries, it enthroned religion in the privacy of the individual soul, not without some sighs of sober satisfaction at its abdication from society. Professor Dicey has commented on the manner in which "the appeal of the Evangelicals to personal religion corresponds with the appeal of Benthamite Liberals to individual energy." [106] The same affinity between religious and social interests found an even clearer expression in the Puritan movement of the seventeenth century. Individualism in religion led insensibly, if not quite logically, to an individualist morality, and an individualist morality to a disparagement of the significance of the social fabric as compared with personal character.

A practical example of that change of emphasis is given by the treatment accorded to the questions of Enclosure and of Pauperism.  For a century and a half the progress of enclosing had been a burning issue, flaring up, from time to time, into acute agitation.  During the greater part of that period, from Latimer in the thirties of the sixteenth century to Laud in the thirties of the seventeenth, the attitude of religious teachers had been one of condemnation.  Sermon after sermon and pamphlet after pamphlet—not to mention Statutes and Royal Commissions—had been launched against depopulation.  The appeal had been, not merely to public policy, but to religion.  Peasant and lord, in their different degrees, are members of one Christian commonwealth, within which the law of charity must bridle the corroding appetite for economic gain.  In such a mystical corporation, knit together by mutual obligations, no man may press his advantage to the full, for no man may seek to live "outside the body of the Church."

Sabotaged by the unpaid magistracy of country gentlemen, who had been the obstructive agents of local administration, the practical application of such doctrines had always been intermittent, and, when the Long Parliament struck the weapon of administrative law from the hands of the Crown, it had ceased altogether.  But the politics of Westminster were not those of village and borough.  The events which seemed to aristocratic Parliamentarians to close the revolution seemed to the left wing of the victorious army only to begin it.  In that earliest and most turbulent of English democracies, where buff-coat taught scripture politics to his general, the talk was not merely of political, but of social, reconstruction.  The program of the Levellers, who more than any other party could claim to express the aspirations of the unprivileged classes, included a demand, not only for annual or biennial Parliaments, manhood suffrage, a redistribution of seats in proportion to population, and the abolition of the veto of the House of Lords, but also that "you would have laid open all enclosures of fens and other commons, or have them enclosed only or chiefly for the benefit of the poor." [107] Theoretical communism, repudiated by the leading Levellers, found its expression in the agitation of the Diggers, on whose behalf Winstanley argued that, "seeing the common people of England, by' joynt consent of person and purse, have caste out Charles, our Norman oppressour . . . the land now is to returne into the joynt hands of those who have conquered, that is the commonours," and that the victory over the King was incomplete, as long as "wee . . . remayne slaves still to the kingly power in the hands of lords of manors." [108]

Nor was it only from the visionary and the zealot that the pressure for redress proceeded. When the shattering of traditional authority seemed for a moment to make all things new, local grievances, buried beneath centuries of dull oppression, started to life, and in several Midland counties the peasants rose to pull down the hated hedges. At Leicester, where in 1649 there were rumors of a popular movement to throw down the enclosures of the neighboring forest, the City Council took the matter up. A petition was drafted, setting out the economic and social evils attending enclosure, and proposing the establishment of machinery to check it, consisting of a committee without whose assent enclosing was not to be permitted. A local minister was instructed to submit the petition to Parliament, "which hath still a watchful eye and open ear to redress the common grievances of the nation." [109] The agent selected to present the city's case was the Rev. John Moore, a prolific pamphleteer, who for several years attacked the depopulating landlord with all the fervor of Latimer, though with even less than Latimer's success.

Half a century before, such commotions would have been followed by the passing of Depopulation Acts and the issue of a Royal Commission.   But, in the ten years since the meeting of the Long Parliament, the whole attitude of public policy towards the movement had begun to change.   Confiscations, compositions and war taxation had effected a revolution in the distribution of property, similar, on a smaller scale, to that which had taken place at the Reformation.   As land changed hands, customary relations were shaken and new interests were created.   Enclosure, as Moore complained,[110] was being pushed forward by means of law suits ending in Chancery decrees.   It was not to be expected that City merchants and members of the Committee for Compounding, some of whom had found land speculation a profitable business, should hear with enthusiasm a proposal to revive the old policy of arresting enclosures by State interference, at which the gentry had grumbled for more than a century.

In these circumstances, it is not surprising that reformers should have found the open ear of Parliament impenetrably closed to agrarian grievances.   Nor was it only the political and economic environment which had changed. The revolution in thought was equally profound.   The theoretical basis of the policy of protecting the peasant by preventing enclosure had been a conception of landownership which regarded its rights and its duties as inextricably interwoven.   Property was not merely a source of income, but a public function, and its use was limited by social obligations and necessities of State.   With such a doctrine the classes who had taken the lead in the struggle against the monarchy could make no truce.   Its last vestiges finally disappeared when the Restoration Parliament swept away military tenures, and imposed on the nation, in the shape of an excise, the financial burden previously borne by themselves.

The theory which took its place, and which was to become in the eighteenth century almost a religion, was that

expressed by Locke, when he described property as a right
anterior to the existence of the State, and argued that "the
supreme power cannot take from any man any part of his
property without his own consent." But Locke merely
poured into a philosophical mould ideas which had been
hammered out in the stress of political struggles, and which
were already the commonplace of landowner and merchant.
The view of society held by that part of the Puritan movement which was socially and politically influential had been
expressed by Ireton and Cromwell in their retort to the
democrats in the army. It was that only the freeholders
really constituted the body politic, and that they could use
their property as they pleased, uncontrolled by obligations
to any superior, or by the need of consulting the mass of
men, who were mere tenants at will, with no fixed interest or
share in the land of the kingdom.[111] Naturally, this change
of ideas had profound reactions on agrarian policy. Formerly a course commending itself to all public-spirited persons, the prevention of enclosure was now discredited as the
program of a sect of religious and political radicals.
When Major-General Whalley in 1656 introduced a measure to regulate and restrict the enclosure of commons,
framed, apparently, on the lines proposed by the authorities
of Leicester, there was an instant outcry from members that
it would "destroy property," and the bill was refused a second reading.[112] After the Restoration the tide began to run
more strongly in the same direction. Enclosure had already
become the hobby of the country gentleman. Experts advo-
cated it on economic grounds, and legislation to facilitate it
was introduced into Parliament. Though its technique still
remained to be elaborated, the attitude which was to be decisive in the eighteenth century had already been crystallized.

The change of policy was striking. The reason of it was
not merely that political conditions made the landed gentry

omnipotent, and that the Royalist squirearchy, who streamed back to their plundered manors in 1660, were in no mood to countenance a revival, by the Government of Charles II, of the administrative interference with the rights of property which had infuriated them in the Government of Charles I. It was that opinion as to social policy had changed, and changed not least among men of religion themselves. The pursuit of economic self-interest, which is the law of nature, is already coming to be identified by the pious with the operation of the providential plan, which is the law of God. Enclosures will increase the output of wool and grain. Each man knows best what his land is suited to produce, and the general interest will be best served by leaving him free to produce it. "It is an undeniable maxim that every one by the light of nature and reason will do that which makes for his greatest advantage. . . . The advancement of private persons will be the advantage of the public." [113]

It is significant that such considerations were adduced, not by an economist, but by a minister. For the argument was ethical as well as economic, and, when Moore appealed to the precepts of traditional morality to bridle pecuniary interests, he provoked the retort that a judicious attention to pecuniary interests was an essential part of an enlightened morality. What the poor need for their spiritual health is—to use the favorite catchword of the age—"regulation," and regulation is possible only if they work under the eye of an employer. In the eyes of the austere moralists of the Restoration, the first, and most neglected, virtue of the poor is industry. Common rights encourage idleness by offering a precarious and demoralizing livelihood to men who ought to be at work for a master. It is not surprising, therefore, that the admonitions of religious teachers against the wickedness of joining house to house and field to field should almost entirely cease. Long the typical example of uncharitable covetousness, enclosure is now considered, not merely economically expedient, but morally beneficial. Baxter, with all his scrupulousness—partly, perhaps, because of his scrupulousness—differs from most earlier divines in giving a qualified approval to enclosure "done in moderation by a pious man," for the characteristic reason that a master can establish a moral discipline among his employees, which they would miss if they worked for themselves. What matters, in short, is not their circumstances, but their character. If they lose as peasants, they will gain as Christians. Opportunities for spiritual edification are more important than the mere material environment. If only the material environment were not itself among the forces determining men's capacity to be edified!

The temper which deplored that the open-field village was not a school of the severer virtues turned on pauperism and poor relief an even more shattering criticism. There is no province of social life in which the fashioning of a new scale of ethical values on the Puritan anvil is more clearly revealed. In the little communities of peasants and craftsmen which composed medieval England, all, when Heaven sent a bad harvest, had starved together, and the misery of the sick, the orphan and the aged had appeared as a personal calamity, not as a social problem. Apart from a few precocious theorists, who hinted at the need for a universal and secular system of provision for distress, the teaching most characteristic of medieval writers had been that the relief of the needy was a primary obligation on those who had means. St. Thomas, who in this matter is typical, quotes with approval the strong words of St. Ambrose about those who cling to the bread of the starving, insists on the idea that property is stewardship, and concludes—a conclusion not always drawn from that well-worn phrase—that to withhold alms when there is evident and urgent necessity is mortal sin.[114] Popular feeling had lent a half-mystical glamour

both to poverty and to the compassion by which poverty
was relieved, for poor men were God's friends.  At best, the
poor were thought to represent our Lord in a peculiarly intimate way—"in that sect," as Langland said, "our Saviour
saved all mankind"—and it was necessary for the author of
a religious manual to explain that the rich, as such, were not
necessarily hateful to God.[115]  At worst, men reflected that
the prayers of the poor availed much, and that the sinner
had been saved from hell by throwing a loaf of bread to a
beggar, even though a curse went with it.  The alms bestowed today would be repaid a thousandfold, when the soul
took its dreadful journey amid rending briars and scorch-
ing flames.

> If ever thou gavest hosen and shoon,
>   *Everie nighte and alle,*
> Sit thee down and put them on,
>   *And Christe receive thy saule.*
>
> If hosen and shoon thou gavest nane,
>   *Everie nighte and alle,*
> The whinnes shall pricke thee to the bare bane,
>   *And Christe receive thy saule.*
>
> .    .    .    .    .
>
> If ever thou gavest meate or drinke,
>   *Everie nighte and alle,*
> The fire shall never make thee shrinke,
>   *And Christe receive thy saule.*
>
> If meate or drinke thou gavest nane,
>   *Everie nighte and alle,*
> The fire will burne thee to the bare bane,
>   *And Christe receive thy saule.*
>
> This ae nighte, this ae nighte,
>   *Everie nighte and alle,*
> Fire, and sleete, and candle-lighte,
>   *And Christe receive thy saule.*[116]

The social character of wealth, which had been the essence of the medieval doctrine, was asserted by English divines in the sixteenth century with redoubled emphasis, precisely because the growing individualism of the age menaced the traditional conception. "The poor man," preached Latimer, "hath title to the rich man's goods; so that the rich man ought to let the poor man have part of his riches to help and to comfort him withal." [117]  Nor had that sovereign indifference to the rigors of the economic calculus disappeared, when, under the influence partly of humanitarian representatives of the Renaissance like Vives, partly of religious reformers, partly of their own ambition to gather all the threads of social administration into their own hands, the statesmen of the sixteenth century set themselves to organize a secular system of poor relief.  In England, after three generations in which the attempt was made to stamp out vagrancy by police measures of hideous brutality, the momentous admission was made that its cause was economic distress, not merely personal idleness, and that the whip had no terrors for the man who must either tramp or starve. The result was the celebrated Acts imposing a compulsory poor-rate and requiring the able-bodied man to be set on work.  The Privy Council, alert to prevent disorder, drove lethargic justices hard, and down to the Civil War the system was administered with fair regularity.  But the Elizabethan Poor Law was never designed to be what, with disastrous results, it became in the eighteenth and early nineteenth centuries, the sole measure for coping with economic distress.  While it provided relief, it was but the last link in a chain of measures—the prevention of evictions, the control of food supplies and prices, the attempt to stabilize employment and to check unnecessary dismissals of workmen—intended to mitigate the forces which made relief necessary. Apart from the Poor Law, the first forty years of the seventeenth century were prolific in the private charity which founded alms-houses and hospitals, and established funds to provide employment or to aid struggling tradesmen. The appeal was still to religion, which owed to poverty a kind of reverence.

> It was Thy choice, whilst Thou on earth didst stay,
> And hadst not whereupon Thy head to lay.[118]

"What, speak you of such things?" said Nicholas Ferrar on his death-bed to one who commended his charities. "It would have been but a suitable return for me to have given all I had, and not to have scattered a few crumbs of alms here and there." [119]

It was inevitable that, in the anarchy of the Civil War, both private charity and public relief should fall on evil days. In London, charitable endowments seem to have suffered from more than ordinary malversation, and there were complaints that the income both of Bridewell and of the Hospitals was seriously reduced.[120] In the country, the records of Quarter Sessions paint a picture of confusion, in which the machinery of presentment by constables to justices has broken down, and a long wail arises, that thieves are multiplied, the poor are neglected, and vagrants wander to and fro at their will.[121] The administrative collapse of the Elizabethan Poor Law continued after the Restoration, and twenty-three years later Sir Matthew Hale complained that the sections in it relating to the provision of employment were a dead letter.[122] Always unpopular with the local authorities, whom they involved in considerable trouble and expense, it is not surprising that, with the cessation of pressure by the Central Government, they should, except here and there, have been neglected. What is more significant, however, than the practical deficiencies in the administration of relief, was the rise of a new school of opinion, which regarded with repugnance the whole body of social theory of which both private charity and public relief had been the expression.

"The generall rule of all England," wrote a pamphleteer in 1646, "is to whip and punish the wandring beggars . . . and so many justices execute one branch of that good Statute (which is the point of justice), but as for the point of charitie, they leave [it] undone, which is to provide houses and convenient places to set the poore to work." [123]    The House of Commons appears to have been conscious that the complaint had some foundation; in 1649 it ordered that the county justices should be required to see that stocks of material were provided as the law required,[124] and the question of preparing new legislation to ensure that persons in distress should be found employment was on several occasions referred to committees of the House.[125]    Nothing seems, however, to have come of these proposals, nor was the Elizabethan policy of "setting the poor on work" that which was most congenial to the temper of the time.    Upon the admission that distress was the result, not of personal deficiencies, but of economic causes, with its corollary that its victims had a legal right to be maintained by society, the growing individualism of the age turned the same frigid scepticism as was later directed against the Speenhamland policy by the reformers of 1834.    Like the friends of Job, it saw in misfortune, not the chastisement of love, but the punishment for sin.    The result was that, while the penalties on the vagrant were redoubled, religious opinion laid less emphasis on the obligation of charity than upon the duty of work, and that the admonitions which had formerly been turned upon uncharitable covetousness were now directed against improvidence and idleness.    The characteristic sentiment was that of Milton's friend, Hartlib: "The law of God saith, 'he that will not work, let him not eat.'    This would be a sore scourge and smart whip for idle persons if . . .

none should be suffered to eat till they had wrought for it." [126]

The new attitude found expression in the rare bursts of public activity provoked by the growth of pauperism between 1640 and 1660. The idea of dealing with it on sound business principles. by means of a corporation which would combine profit with philanthropy, was being sedulously preached by a small group of reformers.[127]  Parliament took it up, and in 1649 passed an Act for the relief and employment of the poor and the punishment of beggars, under which a company was to be established with power to apprehend vagrants, to offer them the choice between work and whipping, and to set to compulsory labor all other poor persons, including children without means of maintenance.[128] Eight years later the prevalence of vagrancy produced an Act of such extreme severity as almost to recall the suggestion made a generation later by Fletcher of Saltoun, that vagrants should be sent to the galleys.  It provided that, since offenders could rarely be taken in the act, any vagrant who failed to satisfy the justices that he had a good reason for being on the roads should be arrested and punished as a sturdy beggar, whether actually begging or not.[129]

The protest against indiscriminate almsgiving, as the parade of a spurious religion, which sacrificed character to a formal piety, was older than the Reformation, but it had been given a new emphasis by the reformers.  Luther had denounced the demands of beggars as blackmail, and the Swiss reformers had stamped out the remnants of monastic charity, as a bribe ministered by Popery to dissoluteness and demoralization.  "I conclude that all the large givings of the papists," preached an English divine in the reign of Elizabeth, "of which at this day many make so great brags, because they be not done in a reverent regard of the commandment of the Lord, in love, and of an inward being touched with the calamities of the needy, but for to be well reported of before men whilst they are alive, and to be prayed for after they are dead . . . are indeed no alms, but pharisaical trumpets." [130]   The rise of a commercial civilization, the reaction against the authoritarian social policy of the Tudors, and the progress of Puritanism among the middle classes, all combined in the next half-century to sharpen the edge of that doctrine.   Nurtured in a tradition which made the discipline of character by industry and self-denial the center of its ethical scheme, the Puritan moralist was undisturbed by any doubts as to whether even the seed of the righteous might not sometimes be constrained to beg its bread, and met the taunt that the repudiation of good works was the cloak for a conscienceless egoism with the retort that the easy-going open-handedness of the sentimentalist was not less selfish in its motives and was more corrupting to its objects.   "As for idle beggars," wrote Steele, "happy for them if fewer people spent their foolish pity upon their bodies, and if more shewed some wise compassion upon their souls." [131]   That the greatest of evils is idleness, that the poor are the victims, not of circumstances, but of their own "idle, irregular and wicked courses," that the truest charity is not to enervate them by relief, but so to reform their characters that relief may be unnecessary—such doctrines turned severity from a sin into a duty, and froze the impulse of natural pity with the assurance that, if indulged, it would perpetuate the suffering which it sought to allay.

Few tricks of the unsophisticated intellect are more curious than the naïve psychology of the business man, who ascribes his achievements to his own unaided efforts, in bland unconsciousness of a social order without whose continuous support and vigilant protection he would be as a lamb bleating in the desert.   That individualist complex owes part of its self-assurance to the suggestion of Puritan moralists, that practical success is at once the sign and the reward of ethical superiority. "No question," argued a Puritan pamphleteer, "but it [riches] should be the portion rather of the godly than of the wicked, were it good for them; for godliness hath the promises of this life as well as of the life to come." [132]  The demonstration that distress is a proof of demerit, though a singular commentary on the lives of Christian saints and sages, has always been popular with the prosperous.  By the lusty plutocracy of the Restoration, roaring after its meat, and not indisposed, if it could not find it elsewhere, to seek it from God, it was welcomed with a shout of applause.

A society which reverences the attainment of riches as the supreme felicity will naturally be disposed to regard the poor as damned in the next world, if only to justify itself for making their life a hell in this.  Advanced by men of religion as a tonic for the soul, the doctrine of the danger of pampering poverty was hailed by the rising school of Political Arithmeticians as a sovereign cure for the ills of society.  For, if the theme of the moralist was that an easygoing indulgence undermined character, the theme of the economist was that it was economically disastrous and financially ruinous.  The Poor Law is the mother of idleness, "men and women growing so idle and proud that they will not work, but lie upon the parish wherein they dwell for maintenance."  It discourages thrift; "if shame or fear of punishment makes him earn his dayly bread, he will do no more; his children are the charge of the parish and his old age his recess from labour or care."  It keeps up wages, since "it encourages wilful and evil-disposed persons to impose what wages they please upon their labours; and herein they are so refractory to reason and the benefit of the nation that, when corn and provisions are cheap, they will not work for less wages than when they were dear." [133]  To the landowner who cursed the poor-rates, and the clothier who grumbled at the high cost of labor, one school of religious thought now brought the comforting assurance that morality itself would be favored by a reduction of both.

As the history of the Poor Law in the nineteenth century was to prove, there is no touchstone, except the treatment of childhood, which reveals the true character of a social philosophy more clearly than the spirit in which it regards the misfortunes of those of its members who fall by the way. Such utterances on the subject of poverty were merely one example of a general attitude, which appeared at times to consign to collective perdition almost the whole of the wage-earning population. It was partly that, in an age which worshiped property as the foundation of the social order, the mere laborer seemed something less than a full citizen. It was partly the result of the greatly increased influence on thought and public affairs acquired at the Restoration by the commercial classes, whose temper was a ruthless materialism, determined at all costs to conquer world-markets from France and Holland, and prepared to sacrifice every other consideration to their economic ambitions. It was partly that, in spite of a century of large-scale production in textiles, the problems of capitalist industry and of a propertyless proletariat were still too novel for their essential features to be appreciated. Even those writers, like Baxter and Bunyan, who continued to insist on the wickedness of extortionate prices and unconscionable interest, rarely thought of applying their principles to the subject of wages. Their social theory had been designed for an age of petty agriculture and industry, in which personal relations had not yet been superseded by the cash nexus, and the craftsman or peasant farmer was but little removed in economic status from the half-dozen journeymen or laborers whom he employed. In a world increasingly dominated by great clothiers, iron-masters and mine-owners, they still adhered to the antiquated categories of master and servant, with the same obstinate indifference to economic realities as leads the twentieth century to talk of employers and employed, long after the individual employer has been converted into an impersonal corporation.

In a famous passage of the *Communist Manifesto*, Marx observes that "the *bourgeoisie*, wherever it got the upper hand, put an end to all feudal, patriarchal, idyllic relations, pitilessly tore asunder the motley feudal ties that bound man to his 'natural superiors,' and left remaining no other bond between man and man than naked self-interest and callous cash payment." [134] An interesting illustration of his thesis might be found in the discussions of the economics of employment by English writers of the period between 1660 and 1760. Their characteristic was an attitude towards the new industrial proletariat noticeably harsher than that general in the first half of the seventeenth century, and which has no modern parallel except in the behavior of the less reputable of white colonists towards colored labor. The denunciations of the "luxury, pride and sloth" [135] of the English wage-earners of the seventeenth and eighteenth centuries are, indeed, almost exactly identical with those directed against African natives today. It is complained that, compared with the Dutch, they are self-indulgent and idle; that they want no more than a bare subsistence, and will cease work the moment they obtain it; that, the higher their wages, the more—"so licentious are they" [136]—they spend upon drink; that high prices, therefore, are not a misfortune, but a blessing, since they compel the wage-earner to be more industrious; and that high wages are not a blessing, but a misfortune, since they merely conduce to "weekly debauches."

When such doctrines were general, it was natural that the rigors of economic exploitation should be preached as a public duty, and, with a few exceptions, the writers of the period differed only as to the methods by which severity could most advantageously be organized. Pollexfen and

Walter Harris thought that salvation might be found by reducing the number of days kept as holidays. Bishop Berkeley, with the conditions of Ireland before his eyes, suggested that "sturdy beggars should . . . be seized and made slaves to the public for a certain term of years." Thomas Alcock, who was shocked at the workman's taste for snuff, tea and ribbons, proposed the revival of sumptuary legislation.[137] The writers who advanced schemes for reformed workhouses, which should be places at once of punishment and of training, were innumerable. All were agreed that, on moral no less than on economic grounds, it was vital that wages should be reduced. The doctrine afterwards expressed by Arthur Young, when he wrote, "every one but an idiot knows that the lower classes must be kept poor, or they will never be industrious," [138] was the tritest commonplace of Restoration economists. It was not argued; it was accepted as self-evident.

When philanthropists were inquiring whether it might not be desirable to reëstablish slavery, it was not to be expected that the sufferings of the destitute would wring their hearts with social compunction. The most curious feature in the whole discussion, and that which is most sharply in contrast with the long debate on pauperism carried on in the sixteenth century, was the resolute refusal to admit that society had any responsibility for the causes of distress. Tudor divines and statesmen had little mercy for idle rogues. But the former always, and the latter ultimately, regarded pauperism primarily as a social phenomenon produced by economic dislocation, and the embarrassing question put by the genial Harrison—"at whose handes shall the bloude of these men be required?" [139]—was never far from the minds even of the most cynical. Their successors after the Restoration were apparently quite unconscious that it was even conceivable that there might be any other cause of poverty than the moral failings of the poor. The practical conclusion to be drawn from so comfortable a creed was at once extremely simple and extremely agreeable. It was not to find employment under the Act of 1601, for to do that was only "to render the poor more bold." It was to surround the right to relief with obstacles such as those contained in the Act of 1662, to give it, when it could not be avoided, in a workhouse or house of correction, and, for the rest, to increase the demand for labor by reducing wages.

The grand discovery of a commercial age, that relief might be so administered as not merely to relieve, but also to deter, still remained to be made by Utilitarian philosophers. But the theory that distress was due, not to economic circumstances, but to what the Poor Law Commissioners of 1834 called "individual improvidence and vice," was firmly established, and the criticism on the Elizabethan system which was to inspire the new Poor Law had already been formulated. The essence of that system was admirably expressed a century later by a Scottish divine as "the principle that each man, simply because he exists, holds a right on other men or on society for existence." [140] Dr. Chalmers' attack upon it was the echo of a note long struck by Puritan moralists. And the views of Dr. Chalmers had impressed themselves on Nassau Senior,[141] before he set his hand to that brilliant, influential and wildly unhistorical Report, which, after provoking something like a rebellion in the north of England, was to be one of the pillars of the social policy of the nineteenth century.

It would be misleading to dwell on the limitations of Puritan ethics without emphasizing the enormous contribution of Puritanism to political freedom and social progress. The foundation of democracy is the sense of spiritual independence which nerves the individual to stand alone against the powers of this world, and in England, where squire and parson, lifting arrogant eyebrows at the insolence of the lower orders, combined to crush popular agitation, as a menace at once to society and to the Church, it is probable that democracy owes more to Nonconformity than to any other single movement. The virtues of enterprise, diligence and thrift are the indispensable foundation of any complex and vigorous civilization. It was Puritanism which, by investing them with a supernatural sanction, turned them from an unsocial eccentricity into a habit and a religion. Nor would it be difficult to find notable representatives of the Puritan spirit in whom the personal austerity, which was the noblest aspect of the new ideal, was combined with a profound consciousness of social solidarity, which was the noblest aspect of that which it displaced. Firmin the philanthropist, and Bellers the Quaker, whom Owen more than a century later hailed as the father of his doctrines, were pioneers of Poor Law reform. The Society of Friends, in an age when the divorce between religion and social ethics was almost complete, met the prevalent doctrine, that it was permissible to take such gain as the market offered, by insisting on the obligation of good conscience and forbearance in economic transactions, and on the duty to make the honorable maintenance of the brother in distress a common charge.[142]

The general climate and character of a country are not altered, however, by the fact that here and there it has peaks which rise into an ampler air. The distinctive note of Puritan teaching was different. It was individual responsibility, not social obligation. Training its pupils to the mastery of others through the mastery of self, it prized as a crown of glory the qualities which arm the spiritual athlete for his solitary contest with a hostile world, and dismissed concern with the social order as the prop of weaklings and the Capua of the soul. Both the excellences and the defects of that attitude were momentous for the future. It is sometimes suggested that the astonishing outburst of industrial activity which took place after 1760 created a new type of economic character, as well as a new system of economic organization. In reality, the ideal which was later to carry all before it, in the person of the inventor and engineer and captain of industry, was well established among Englishmen before the end of the seventeenth century. Among the numerous forces which had gone to form it, some not inconsiderable part may reasonably be ascribed to the emphasis on the life of business enterprise as the appropriate field for Christian endeavor, and on the qualities needed for success in it, which was characteristic of Puritanism. These qualities, and the admiration of them, remained, when the religious reference, and the restraints which it imposed, had weakened or disappeared.

# CHAPTER V

## CONCLUSION

"Ther is a certaine man that shortly after my fyrst sermon, beynge asked if he had bene at the sermon that day, answered, yea. I praye you, said he, how lyked you hym? Mary, sayed he, even as I lyked hym alwayes—a sedicious fellow."

LATIMER, *Seven Sermons before King Edward VI.*

# CHAPTER V

SOCIETIES, like individuals, have their moral crises and their spiritual revolutions. The student can observe the results which these cataclysms produce, but he can hardly without presumption attempt to appraise them, for it is at the fire which they kindled that his own small taper has been lit. The rise of a naturalistic science of society, with all its magnificent promise of fruitful action and of intellectual light; the abdication of the Christian Churches from departments of economic conduct and social theory long claimed as their province; the general acceptance by thinkers of a scale of ethical values, which turned the desire for pecuniary gain from a perilous, if natural, frailty into the idol of philosophers and the mainspring of society—such movements are written large over the history of the tempestuous age which lies between the Reformation and the full light of the eighteenth century. Their consequences have been worked into the very tissue of modern civilization. Posterity still stands too near their source to discern the ocean into which these streams will flow.

In an historical age the relativity of political doctrines is the tritest of commonplaces. But social psychology continues too often to be discussed in serene indifference to the categories of time and place, and economic interests are still popularly treated as though they formed a kingdom over which the *Zeitgeist* bears no sway. In reality, though inherited dispositions may be constant from generation to generation, the system of valuations, preferences and ideals —the social environment within which individual character

277

functions—is in process of continuous change, and it is in the conception of the place to be assigned to economic interests in the life of society that change has in recent centuries been most comprehensive in its scope, and most sensational in its consequences. The isolation of economic aims as a specialized object of concentrated and systematic effort, the erection of economic criteria into an independent and authoritative standard of social expediency, are phenomena which, though familiar enough in classical antiquity, appear, at least on a grand scale, only at a comparatively recent date in the history of later civilizations. The conflict between the economic outlook of East and West, which impresses the traveller today, finds a parallel in the contrast between medieval and modern economic ideas, which strikes the historian.

The elements which combined to produce that revolution are too numerous to be summarized in any neat formula. But, side by side with the expansion of trade and the rise of new classes to political power, there was a further cause, which, if not the most conspicuous, was not the least fundamental. It was the contraction of the territory within which the spirit of religion was conceived to run. The criticism which dismisses the concern of Churches with economic relations and social organization as a modern innovation finds little support in past history. What requires explanation is not the view that these matters are part of the province of religion, but the view that they are not. When the age of the Reformation begins, economics is still a branch of ethics, and ethics of theology; all human activities are treated as falling within a single scheme, whose character is determined by the spiritual destiny of mankind; the appeal of theorists is to natural law, not to utility; the legitimacy of economic transactions is tried by reference, less to the movements of the market, than to moral standards derived from the traditional teaching of the Christian

Church; the Church itself is regarded as a society wielding theoretical, and sometimes practical, authority in social affairs. The secularization of political thought, which was to be the work of the next two centuries, had profound reactions on social speculation, and by the Restoration the whole perspective, at least in England, has been revolutionized. Religion has been converted from the keystone which holds together the social edifice into one department within it, and the idea of a rule of right is replaced by economic expediency as the arbiter of policy and the criterion of conduct. From a spiritual being, who, in order to survive, must devote a reasonable attention to economic interest, man seems sometimes to have become an economic animal, who will be prudent, nevertheless, if he takes due precautions to assure his spiritual well-being.

The result is an attitude which forms so fundamental a part of modern political thought, that both its precarious philosophical basis, and the contrast which it offers with the conceptions of earlier generations, are commonly forgotten. Its essence is a dualism which regards the secular and the religious aspects of life, not as successive stages within a larger unity, but as parallel and independent provinces, governed by different laws, judged by different standards, and amenable to different authorities. To the most representative minds of the Reformation, as of the Middle Ages, a philosophy which treated the transactions of commerce and the institutions of society as indifferent to religion would have appeared, not merely morally reprehensible, but intellectually absurd. Holding as their first assumption that the ultimate social authority is the will of God, and that temporal interests are a transitory episode in the life of spirits which are eternal, they state the rules to which the social conduct of the Christian must conform, and, when circumstances allow, organize the discipline by which those rules may be enforced. By their successors in the eighteenth cen-

tury the philosophy of Indifferentism, though rarely formulated as a matter of theory, is held in practice as a truism
which it is irrational, if not actually immoral, to question,
since it is in the heart of the individual that religion has its
throne, and to externalize it in rules and institutions is to
tarnish its purity and to degrade its appeal. Naturally,
therefore, they formulate the ethical principles of Christianity in terms of a comfortable ambiguity, and rarely indicate
with any precision their application to commerce, finance,
and the ownership of property. Thus the conflict between
religion and those natural economic ambitions which the
thought of an earlier age had regarded with suspicion is
suspended by a truce which divides the life of mankind
between them. The former takes as its province the individual soul, the latter the intercourse of man with his
fellows in the activities of business and the affairs of society.
Provided that each keeps to its own territory, peace is
assured. They cannot collide, for they can never meet.

History is a stage where forces which are within human
control contend and coöperate with forces which are not.
The change of opinion described in these pages drew nourishment from both. The storm and fury of the Puritan
revolution had been followed by a dazzling outburst of
economic enterprise, and the transformation of the material
environment prepared an atmosphere in which a judicious
moderation seemed the voice at once of the truest wisdom
and the sincerest piety. But the inner world was in motion
as well as the outer. The march of external progress woke
sympathetic echoes in hearts already attuned to applaud its
triumph, and there was no consciousness of an acute tension
between the claims of religion and the glittering allurements
of a commercial civilization, such as had tormented the age
of the Reformation.

It was partly the natural, and not unreasonable, diffidence
of men who were conscious that traditional doctrines of

social ethics, with their impracticable distrust of economic motives, belonged to the conditions of a vanished age, but who lacked the creative energy to state them anew, in a form applicable to the needs of a more complex and mobile social order. It was partly that political changes had gone far to identify the Church of England with the ruling aristocracy, so that, while in France, when the crash came, many of the lower clergy threw in their lot with the *tiers état,* in England it was rarely that the officers of the Church did not echo the views of society which commended themselves to the rulers of the State. It was partly that, to one important body of opinion, the very heart of religion was a spirit which made indifference to the gross world of external circumstances appear, not a defect, but an ornament of the soul. Untrammelled by the silken chains which bound the Establishment, and with a great tradition of discipline behind them, the Nonconformist Churches might seem to have possessed opportunities of reasserting the social obligations of religion with a vigor denied to the Church of England. What impeded their utterance was less a weakness than the most essential and distinctive of their virtues. Founded on the repudiation of the idea that human effort could avail to win salvation, or human aid to assist the pilgrim in his lonely quest, they saw the world of business and society as a battlefield, across which character could march triumphant to its goal, not as crude materials waiting the architect's hand to set them in their place as the foundations of the Kingdom of Heaven. It did not occur to them that character is social, and society, since it is the expression of character, spiritual. Thus the eye is sometimes blinded by light itself.

The certainties of one age are the problems of the next. Few will refuse their admiration to the magnificent conception of a community penetrated from apex to foundation by the moral law, which was the inspiration of the great reformers, not less than of the better minds of the Middle Ages. But, in order to subdue the tough world of material interests, it is necessary to have at least so much sympathy with its tortuous ways as is needed to understand them. The Prince of Darkness has a right to a courteous hearing and a fair trial, and those who will not give him his due are wont to find that, in the long run, he turns the tables by taking his due and something over. Common sense and a respect for realities are not less graces of the spirit than moral zeal. The paroxysms of virtuous fury, with which the children of light denounced each new victory of economic enterprise as yet another stratagem of Mammon, disabled them for the staff-work of their campaign, which needs a cool head as well as a stout heart. Their obstinate refusal to revise old formulæ in the light of new facts exposed them helpless to a counter-attack, in which the whole fabric of their philosophy, truth and fantasy alike, was overwhelmed together. They despised knowledge, and knowledge destroyed them.

Few can contemplate without a sense of exhilaration the splendid achievements of practical energy and technical skill, which, from the latter part of the seventeenth century, were transforming the face of material civilization, and of which England was the daring, if not too scrupulous, pioneer. If, however, economic ambitions are good servants, they are bad masters. Harnessed to a social purpose, they will turn the mill and grind the corn. But the question, to what end the wheels revolve, still remains; and on that question the naïve and uncritical worship of economic power, which is the mood of unreason too often engendered in those whom that new Leviathan has hypnotized by its spell, throws no light. Its result is not seldom a world in which men commands a mechanism that they cannot fully use, and an organization which has every perfection except that of motion.

*Er nennt's Vernunft und braucht's allein,*
*Nur tierischer als jedes Tier zu sein.*

The shaft of Mephistopheles, which drops harmless from
the armor of Reason, pierces the lazy caricature which
masquerades beneath that sacred name, to flatter its followers with the smiling illusion of progress won from the
mastery of the material environment by a race too selfish
and superficial to determine the purpose to which its
triumphs shall be applied.  Mankind may wring her secrets
from nature, and use their knowledge to destroy themselves; they may command the Ariels of heat and motion,
and bind their wings in helpless frustration, while they
wrangle over the question of the master whom the imprisoned genii shall serve.  Whether the chemist shall provide them with the means of life or with tri-nitro-toluol
and poison gas, whether industry shall straighten the bent
back or crush it beneath heavier burdens, depends on an act
of choice between incompatible ideals, for which no increase
in the apparatus of civilization at man's disposal is in itself
a substitute.  Economic efficiency is a necessary element in
the life of any sane and vigorous society, and only the incor-
rigible sentimentalist will depreciate its significance.  But to
convert efficiency from an instrument into a primary object
is to destroy efficiency itself.  For the condition of effective
action in a complex civilization is coöperation.  And the
condition of coöperation is agreement, both as to the ends to
which effort should be applied, and the criteria by which its
success is to be judged.

Agreement as to ends implies the acceptance of a standard of values, by which the position to be assigned to different objects may be determined.  In a world of limited
resources, where nature yields a return only to prolonged
and systematic effort, such a standard must obviously take
account of economic possibilities.  But it cannot itself be

merely economic, since the comparative importance of economic and of other interests—the sacrifice, for example, of material goods worth incurring in order to extend leisure, or develop education, or humanize toil—is precisely the point on which it is needed to throw light.  It must be based on some conception of the requirements of human nature as a whole, to which the satisfaction of economic needs is evidently vital, but which demands the satisfaction of other needs as well, and which can organize its activities on a rational system only in so far as it has a clear apprehension of their relative significance.  "Whatever the world thinks," wrote Bishop Berkeley, "he who hath not much meditated upon God, the human mind and the *summum bonum* may possibly make a thriving earthworm, but will most indubitably make a sorry patriot and a sorry statesman." The philosopher of today, who bids us base our hopes of progress on knowledge inspired by love, does not differ from the Bishop so much, perhaps, as he would wish.  The most obvious facts are the most easily forgotten.  Both the existing economic order, and too many of the projects advanced for reconstructing it, break down through their neglect of the truism that, since even quite common men have souls, no increase in material wealth will compensate them for arrangements which insult their self-respect and impair their freedom.  A reasonable estimate of economic organization must allow for the fact that, unless industry is to be paralyzed by recurrent revolts on the part of outraged human nature, it must satisfy criteria which are not purely economic.  A reasonable view of its possible modifications must recognize that natural appetites may be purified or restrained, as, in fact, in some considerable measure they already have been, by being submitted to the control of some larger body of interests.  The distinction made by the philosophers of classical antiquity between liberal and servile occupations, the medieval insistence that riches exist for man, not man for riches, Ruskin's famous outburst, "there is no wealth but life," the argument of the Socialist who urges that production should be organized for service, not for profit, are but different attempts to emphasize the instrumental character of economic activities by reference to an ideal which is held to express the true nature of man.

Of that nature and its possibilities the Christian Church was thought, during the greater part of the period discussed in these pages, to hold by definition a conception distinctively its own. It was therefore committed to the formulation of a social theory, not as a philanthropic gloss upon the main body of its teaching, but as a vital element in a creed concerned with the destiny of men whose character is formed, and whose spiritual potentialities are fostered or starved, by the commerce of the market-place and the institutions of society. Stripped of the eccentricities of period and place, its philosophy had as its center a determination to assert the superiority of moral principles over economic appetites, which have their place, and an important place, in the human scheme, but which, like other natural appetites, when flattered and pampered and overfed, bring ruin to the soul and confusion to society. Its casuistry was an attempt to translate these principles into a code of practical ethics, sufficiently precise to be applied to the dusty world of warehouse and farm. Its discipline was an effort, too often corrupt and pettifogging in practice, but not ignoble in conception, to work the Christian virtues into the spotted texture of individual character and social conduct. That practice was often a sorry parody on theory is a truism which should need no emphasis. But in a world where principles and conduct are unequally mated, men are to be judged by their reach as well as by their grasp—by the ends at which they aim as well as by the success with which they attain them. The prudent critic will try himself by his achievement rather than by his ideals, and his neighbors, living and dead alike, by their ideals not less than by their achievement.

Circumstances alter from age to age, and the practical interpretation of moral principles must alter with them. Few who consider dispassionately the facts of social history will be disposed to deny that the exploitation of the weak by the powerful, organized for purposes of economic gain, buttressed by imposing systems of law, and screened by decorous draperies of virtuous sentiment and resounding rhetoric, has been a permanent feature in the life of most communities that the world has yet seen. But the quality in modern societies which is most sharply opposed to the teaching ascribed to the Founder of the Christian Faith lies deeper than the exceptional failures and abnormal follies against which criticism is most commonly directed. It consists in the assumption, accepted by most reformers with hardly less *naïveté* than by the defenders of the established order, that the attainment of material riches is the supreme object of human endeavor and the final criterion of human success. Such a philosophy, plausible, militant, and not indisposed, when hard pressed, to silence criticism by persecution, may triumph or may decline. What is certain is that it is the negation of any system of thought or morals which can, except by a metaphor, be described as Christian. Compromise is as impossible between the Church of Christ and the idolatry of wealth, which is the practical religion of capitalist societies, as it was between the Church and the State idolatry of the Roman Empire.

"Modern capitalism," writes Mr. Keynes, "is absolutely irreligious, without internal union, without much public spirit, often, though not always, a mere congeries of possessors and pursuers." It is that whole system of appetites and values, with its deification of the life of snatching to hoard, and hoarding to snatch, which now, in the hour of its triumph, while the plaudits of the crowd still ring in the

ears of the gladiators and the laurels are still unfaded on
their brows, seems sometimes to leave a taste as of ashes
on the lips of a civilization which has brought to the conquest of its material environment resources unknown in
earlier ages, but which has not yet learned to master itself.
It was against that system, while still in its supple and in-
sinuating youth, before success had caused it to throw aside
the mask of innocence, and while its true nature was unknown even to itself, that the saints and sages of earlier
ages launched their warnings and their denunciations.   The
language in which theologians and preachers expressed their
horror of the sin of covetousness may appear to the modern
reader too murkily sulphurous; their precepts on the contracts of business and the disposition of property may seem
an impracticable pedantry.   But rashness is a more agreeable failing than cowardice, and, when to speak is unpopular,
it is less pardonable to be silent than to say too much.   Posterity has, perhaps, as much to learn from the whirlwind
eloquence with which Latimer scourged injustice and oppression as from the sober respectability of the judicious
Paley—who himself, since there are depths below depths,
was regarded as a dangerous revolutionary by George III.

# NOTES

## PREFACE TO 1937 EDITION

1 References to some of the earlier literature will be found in the notes on subsequent chapters. The following list of recent books and articles is not exhaustive, but it may be of some use to those interested in the subject. E. Troeltsch, *The Social Teaching of the Christian Churches*, 2 vols., London, 1931 (Eng. trans. by Olive Wyon of his *Die Soziallehren der Christlichen Kirchen und Gruppen*, 1912) ; Max Weber, *The Protestant Ethic and the Spirit of Capitalism*, London, 1930 (Eng. trans. by Talcott Parsons of *Die Protestantische Ethik und der Geist des Kapitalismus* in "Archiv fur Sozialwissenschaft und Sozialpolitik," vols. xx (1904) and xxi (1905) ; later reprinted in *Gesammelte Aufsarze zur Religions-soziologie*, 3 vols., Tubingen, 1921) ; H. Hauser, *Les débuts du Capitalisme*, Paris, 1927, chap. ii ("Les Idées économiques de Calvin") ; B. Groethuysen, *Origines de l'esprit bourgeois en France*, Paris, 1927 ; Margaret James, *Social Problems and Policy during the Puritan Revolution*, 1640-1660, London, 1930; Isabel Grubb, *Quakerism and Industry before 1800*, London, 1930; W. J. Warner, *The Wesleyan Movement in the Industrial Revolution*, London, 1930; R. Pascal, *The Social Basis of the German Reformation*, London, 1933; H. M. Robertson, *The Rise of Economic Individualism*, Cambridge, 1933; A. Fanfani, *Le Origini dello Spirito Capitalistico in Italia*, Milan, 1933, and *Cattolicismo e Protestantesimo nella Formazione Storica del Capitalismo*, Milan, 1934 (Eng. trans. *Catholicism, Protestantism and Capitalism*, London, 1935) ; J. Brodrick, S. J., *The Economic Morals of the Jesuits*, London, 1934; E. D. Bebb, *Nonconformity and Social and Economic Life, 1660-1800*, London, 1935. The articles include the following: M. Halbwachs, "Les Origines Puritaines due Capitalisme Moderne" (*Revue d'histoire et de Philosophie réligieuses*, March-April, 1925) and "Économistes et Historiens, Max Weber, une vie, un œuvre" (*Annales d'Histoire Économique et Sociale*, No. 1, 1929) ; H. Sée, "Dans quelle mesure Puritains et Juifs ont-ils contribué au Progrés due Capitalisme Moderne?" (*Revue Historique*, t. CLV, 1927) ; Kemper Fullerton, "Calvinism and Capitalism" (*Harvard Theological Review*, July 1928) ; F. H. Knight, "Historical and Theoretical Issues in the Problem of Modern Capitalism" (*Journal of Economic and Business History*, November 1928) ; Talcott Parsons, "Capitalism in Recent German Literature" (*Journal of Political Economy*, December 1928 and February 1929) ; P. C. Gordon Walker, "Capitalism and the Reformation" (*Economic History Review*, November 1937).

2 For Weber's life and personality, see Marianne Weber, *Max Weber, ein Lebensbild*, Tubingen, 1926, and Karl Jaspers, *Max Weber, Deutsches*

*Wesen im politischen Denken, im Forschen und Philosophieren,* Oldenburg, 1932.

[3] Max Weber, *The Protestant Ethic and the Spirit of Capitalism,* Eng. trans., p. 183.

[4] H. M. Robertson, *Aspects of the Rise of Economic Individualism,* p. xii.

[5] Weber, *op. cit.,* p. 26.

[6] Weber, *op. cit.,* p. 183.

[7] *Ibid.,* p. 183, and note 118 on chap. v: "it would have been easy to proceed . . . to a regular construction which logically deduced everything characteristic of modern culture from Protestant nationalism. But that sort of thing may be left to the type of dilettante who believes in the unity of the group mind and its reducibility to a single formula." "Spiritual" is my rendering of the almost untranslatable *"spiritualistische kausale."*

[8] See below, note 32 on chap. iv, pp. 247-8, and Max Weber, *op. cit.,* pp. 3-11.

[9] Weber, *op. cit.,* 197-8. A chapter expanding the same criticism is contained in H. M. Robertson, *Aspects of the Rise of Economic Individualism,* pp. 57-87. The best treatment of the subject is that of Brentano, *Die Anfange des modernen Kapitalismus,* 1916, pp. 117-57, and *Der Wirtschaftende Mensch in der Geschichte,* Leipzig, 1923, pp. 363 *sq.*

[10] See H. M. Robertson, *op. cit.,* pp. 88-110 and 133-67; and J. Brodrick, S. J., *The Economic Morals of the Jesuits,* which, in addition to correcting Robertson's errors, contains the best account of the economic teaching of the Jesuits available in English.

[11] *E.g.,* H. Wiskemann, *Darstellung der in Deutschland zur Zeit der Reformation herrschenden Nationalokonomischen Ansichten,* Leipzig, 1861; F. Engels, *Socialism, Utopian and Scientific,* London, 1892, Introduction; Alfred Marshall, *Principles of Economics,* 1898, chap. III; W. Cunningham, *Christianity and Social Questions,* London, 1910 (see below, note 33 on chap. iv). The last work, though published seven years after the appearance of Weber's articles, does not refer to them, nor is its argument similar to theirs.

[12] *E.g.,* H. M. Robertson, *op. cit.,* p. xi. "Many writers have taken advantage of an unpopularity of Capitalism in the twentieth century to employ them [*sc.* the theories ascribed to Weber] in attacks on Calvinism, or on other branches of religion." The only Guy Fawkes of the gang—apart, of course, from myself—detected by Mr. Robertson actually firing the train appears to be that implacable incendiary, Mr. Aldous Huxley. "Infected," like the arch-conspirator, Weber, "with a deep hatred of Capitalism," we stand with him condemned of "a general tendency to undermine the basis of Capitalist society" (*ibid.,* pp. 207-8). The guilty secret is out at last.

[13] H. Pirenne, *Les Périodes de l'Histoire Sociale due Capitalisme,* 1914.

# CHAPTER I

[1] J. B. Say, *Cours complet d'Economie politique pratique,* vol. vi, 1829, pp. 351-2.

[2] R. Torrens, *An Essay on the Production of Wealth,* 1821, Preface, p. xiii.

[3] Lloyd George at Portmadoc (*Times,* June 16, 1921).

[4] J. A. Froude, *Revival of Romanism,* in *Short Studies on Great Subjects,* 3rd ser., 1877, p. 108.

[5] J. N. Figgis, *From Gerson to Grotius,* 1916, pp. 21 *seqq.*

[6] Locke, *Two Treatises of Government,* bk. ii, chap. ix, § 124.

[7] Nicholas Oresme, *c.*1320-82, Bishop of Lisieux from 1377. His *Tractatus de origine, natura, jure et mutationibus monetarum* was probably written about 1360. The Latin and French texts have been edited by Wolowski (Paris, 1864), and extracts are translated by A. E. Monroe, *Early Economic Thought,* 1924, pp. 81-102. Its significance is discussed shortly by Cunningham, *Growth of English Industry and Commerce, Early and Middle Ages* (4th ed., 1905, pp. 354-9), and by Wolowski in his introduction. The date of the *De Usuris* of Laurentius de Rodolfis was 1403; a short account of his theories as to the exchanges will be found in E. Schreiber, *Die volkswirthschaftlichen Anschauungen der Scholastik seit Thomas v. Aquin,* 1913, pp. 211-17. The most important works of St. Antonino (1389-1459, Archbishop of Florence, 1446) are the *Summa Theologica, Summa Confessionalis,* and *De Usuris.* Some account of his teaching is given by Carl Ilgner, *Die volkswirthschaftlichen Anschauungen Antonins von Florenz,* 1904; Schreiber, *op. cit.,* pp. 217-23; and Bede Jarrett, *St. Antonino and Mediæval Economics,* 1914. The full title of Baxter's work is *A Christian Directory: a Summ of Practical Theologie and Cases of Conscience.*

[8] See Chap. IV, p. 206.

[9] Benvenuto da Imola, *Comentum super Dantis Comœdiam* (ed. Lacaita), vol. i, p. 579: "Qui facit usuram vadit ad infernum; qui non facit vadit ad inopiam" (quoted by G. G. Coulton, *Social Life in Britain from the Conquest to the Reformation,* 1919, p. 342).

[10] Lanfranc, *Elucidarium,* lib. ii, p. 18 (in *Opera,* ed. J. A. Giles). See also *Vita Sancti Guidonis* (*Bollandists' Acta Sanctorum,* September, vol. iv, p. 43): "Mercatura raro aut nunquam ab aliquo diu sine crimine exerceri potuit."

[11] B. L. Manning, *The People's Faith in the Time of Wyclif,* 1919, p. 186.

[12] Aquinas, *Summa Theologica,* 2a 2æ, div. I, Q. iii, art. viii.

[13] *Ibid.,* 1a 2æ, div. i, Q. xciv, art. ii.

[14] The Bull *Unam Sanctam* of Boniface VIII.

[15] John of Salisbury, *Polycraticus* (ed. C. C. I. Webb), lib. v, cap. ii ("Est autem res publica, sicut Plutarco placet, corpus quoddam quod divini muneris beneficio animatur"), and lib. vi, cap. x, where the analogy is worked out in detail. For Henry VIII's chaplain see Starkey, *A Dialogue between Cardinal Pole and Thomas Lupset* (Early English Text Society, Extra Ser., no. xxxii, 1878).

[16] Chaucer, *The Persone's Tale*, § 66.

[17] *On the Seven Deadly Sins*, chap. xix (*Select English Works of John Wyclif*, ed. T. Arnold, vol. iii, 1871, p. 145).

[18] John of Salisbury, *op. cit.*, lib. vi, cap. x: "Tunc autem totius rei publicæ salus incolumis præclaraque erit, si superiora membra se impendant inferioribus et inferiora superioribus pari jure respondeant, ut singula sint quasi aliorum ad invicem membra."

[19] Wyclif, *op. cit.*, chaps. ix, x, xi, xvii, *passim* (*Works of Wyclif*, ed. T. Arnold, vol. iii, pp. 130, 131, 132, 134, 143).

[20] See, *e.g.*, A. Doren, *Studien aus der Florentiner Wirthschaftsgeschichte*, 1901, vol. i, chaps. v, vii. His final verdict (p. 458) is: "Man kann es getrost aussprechen: es gibt wohl keine Periode in der Weltgeschichte, in der die natürliche Uebermacht des Kapitals über die besitz- und kapitallose Handarbeit rücksichtsloser, freier von sittlichen und rechtlichen Bedenken, naiver in ihrer selbstverständlichen Konsequenz gewaltet hätte, und bis in die entferntesten Folgen zur Geltung gebracht worden wäre, als in der Blütezeit der Florentiner Tuchindustrie." The picture drawn by Pirenne of the textile industry in Flanders (*Belgian Democracy: Its Early History*, trans. by J. V. Saunders, 1915, pp. 128-34) is somewhat similar.

[21] In Jan. 1298/9 there was held a "parliament of carpenters at Milehende, where they bound themselves by a corporal oath not to observe a certain ordinance or provision made by the Mayor and Aldermen touching their craft," and in the following March a "parliament of smiths" was formed, with a common chest (*Calendar of Early Mayor's Court Rolls of the City of London, 1298–1307*, ed. A. H. Thomas, 1924, pp. 25, 33-4).

[22] The figures for Paris are the estimate of Martin Saint-Léon (*Histoire des Corporations de Métiers*, 3rd ed., 1922, pp. 219-20, 224, 226); those for Frankfurt are given by Bücher (*Die Bevölkerung von Frankfurt am Main im XIV und XV Jahrhundert*, 1886, pp. 103, 146, 605). They do not include apprentices, and must not be pressed too far. The conclusion of Martin Saint-Léon is: "Il est certain qu'au moyen âge (abstraction faite des villes de Flandre) il n'existait pas encore un prolétariat, le nombre des ouvriers ne dépassant guère ou n'atteignant même pas celui des maîtres" (*op. cit.*, p. 227 n.). The towns of Italy should be added, as an exception, to those of Flanders, and in any case the statement is not generally true of the later Middle Ages, when there was certainly a wage-earning proletariat in Germany also (see Lamprecht, *Zum Verständnis der wirthschaftlichen und sozialen Wandlungen in Deutschland vom 14. zum 16. Jahrhundert*, in the *Zeitschrift für Sozial- und Wirthschaftsgeschichte*, vol. i, 1893, pp. 191-263), and even, though on a smaller scale, in England.

[23] *The Grete Sentence of Curs Expouned,* chap. xxviii (*Select English Works of Wyclif,* ed. T. Arnold, vol. iii, p. 333). The passage contains comprehensive denunciations of all sorts of combination, in particular, gilds, "men of sutel craft, as fre masons and othere," and "marchauntis, groceris, and vitileris" who "conspiren wickidly togidre that noon of hem schal bie over a certeyn pris, though the thing that thei bien be moche more worthi" (*ibid.,* pp. 333, 334).

Wyclif's argument is of great interest and importance. It is (1), that such associations for mutual aid are unnecessary. No special institutions are needed to promote fraternity, since, quite apart from them, all members of the community are bound to help each other: "Alle the goodnes that is in thes gildes eche man owith for to do bi comyn fraternyte of Cristendom, by Goddis comaundement." (2) That combinations are a conspiracy against the public. Both doctrines were points in the case for the sovereignty of the unitary State, and both were to play a large part in subsequent history. They were used by the absolutist statesmen of the sixteenth century as an argument for State control over industry, in place of the obstructive torpor of gilds and boroughs, and by the individualists of the eighteenth century as an argument for free competition. The line of thought as to the relation of minor associations to the State runs from Wyclif to Turgot, Rousseau, Adam Smith, the Act of the Legislative Assembly .in 1792 forbidding trade unions ("Les citoyens de même état ou profession, les ouvriers et compagnons d'un art quelconque ne pourront . . . former des règlements sur leurs prétendus intérêts communs"), and the English Combination Acts.

[24] *Kayser Sigmunds Reformation aller Ständen des Heiligen Römischen Reichs,* printed by Goldast, *Collectio Constitutionum Imperialium,* 1713, vol. iv, pp. 170-200. Its probable date appears to be about 1437. It is discussed shortly by J. S. Schapiro, *Social Reform and the Reformation,* 1909, pp. 93-9.

[25] Martin Saint-Léon, *op. cit.,* p. 187. The author's remark is made *à propos* of a ruling of 1270, fixing minimum rates for textile workers in Paris. It appears, however, to be unduly optimistic. The fact that minimum rates were fixed for textile workers must not be taken as evidence that that policy was common, for in England, and probably in France, the textile trades received special treatment, and minimum rates were fixed for them, while maximum rates were fixed for other, and much more numerous, bodies of workers. What is true is that the medieval assumption with regard to wages, as with regard to the much more important question of prices, was that it was possible to bring them into an agreement with an objective standard of equity, which did not reflect the mere play of economic forces.

[26] "The Cardinals' Gospel," translated from the *Carmina Burana* by G. G. Coulton, in *A Mediæval Garner,* 1910, p. 347.

[27] Printed from the *Carmina Burana* by S. Gaselee, *An Anthology of Medieval Latin,* 1925, pp. 58-9.

[28] Innocent IV gave them in 1248 the title of "Romanæ ecclesiæ filii speciales" (Ehrenberg, *Das Zeitalter der Fugger,* 1896, vol. ii, p. 66).

[29] For Grosstête see Matthew Paris, *Chronica Majora,* vol. v, pp.

404-5 (where he is reported as denouncing the Cahorsines, "whom in our time the holy fathers and teachers . . . had driven out of France, but who have been encouraged and protected by the Pope in England, which did not formerly suffer from this pestilence"), and F. S. Stevenson, *Robert Grosseteste, Bishop of Lincoln,* 1899, pp. 101-4. For the bishop of London and the Cahorsines see Matthew Paris, *Chron. Maj.,* vol. iii, pp. 331-2. A useful collection of references on the whole subject is given by Ehrenberg, *op. cit.,* vol. ii, pp. 64-8.

[30] *Registrum Epistolarum J.'Peckham,* vol. i, p. 18, July 1279 (translated by Coulton, *Social Life in Britain from the Conquest to the Reformation,* p. 345).

[31] For cases of clerical usury see Selden Society, vol. v, 1891, *Leet Jurisdiction in the City of Norwich,* ed. W. Hudson, p. 35; *Hist. MSS. Comm., MSS. of the Marquis of Lothian,* 1905, p. 26; and Th. Bonnin, *Regestrum Visitationum Odonis Rigaldi,* 1852, p. 35. See also note 88 (below).

[32] The Chapter of Notre-Dame appears to have lent money at interest to the citizens of Paris (A. Luchaire, *Social France at the time of Philip Augustus,* translated by E. B. Krehbiel, 1912, p. 130). For the bishop's advice to the usurer see *ibid.,* p. 166.

[33] From a letter of St. Bernard, *c.*1125, printed by Coulton, *A Mediæval Garner,* pp. 68-73.

[34] Aquinas, *De Regimine Principum,* lib. ii, cap. i-vii, where the economic foundations of a State are discussed.

[35] Aquinas, *Summa Theol.,* 2ª 2æ, Q. lxxxiii, art. vi. For St. Antonino's remarks to the same purpose, see Jarrett, *St. Antonino and Mediæval Economics,* p. 59.

[36] Gratian, *Decretum,* pt. ii, causa xii, Q. i, c. ii, § 1.

[37] A good account of St. Antonino's theory of property is given by Ilgner, *Die Volkswirthschaftlichen Anschauungen Antonins von Florenz,* chap. x.

[38] "Sed si esset bonus legislator in patria indigente, deberet locare pro pretio magno huiusmodi mercatores . . . et non tantum eis et familiæ sustentationem necessariam invenire, sed etiam industriam, peritiam, et pericula omnia locare; ergo etiam hoc possunt ipsi in vendendo" (quoted Schreiber, *Die volkswirtschaftlichen Anschauungen der Scholastik seit Thomas v. Aquin,* p. 154).

[39] Henry of Ghent, *Aurea Quodlibeta,* p. 42*b* (quoted Schreiber, *op. cit.,* p. 135).

[40] Gratian, *Decretum,* pt. I, dist. lxxxviii, cap. xi.

[41] Aquinas, *Summa Theol.,* 2ª 2æ, Q. lxxvii, art. iv.

[42] *Ibid.* Trade is unobjectionable, "cum aliquis negotiationi intendit propter publicam utilitatem, ne scilicet res necessariæ ad vitam patriæ desint, et lucrum expetit, non quasi finem, sed quasi stipendium laboris."

[43] Henry of Langenstein, *Tractatus bipartitus de contractibus emptionis et venditionis,* i, 12 (quoted Schreiber, *op. cit.,* p. 197).

[44] See Chap. II, § ii.

[45] Examples of these stories are printed by Coulton, *A Mediæval Garner,* 1910, pp. 212-15, 298, and *Social Life in England from the Conquest to the Reformation,* 1919, p. 346.

⁴⁶ The facts are given by Arturo Segre, *Storia del Commercio,* vol. i, p. 223. For a fuller account of credit and money-lending in Florence, see Doren, *Studien aus der Florentiner Wirthschaftsgeschichte,* vol. i, pp. 173-209.

⁴⁷ Bruno Kuske, *Quellen zur Geschichte des Kölner Handels und Verkehrs im Mittelalter,* vol. iii, 1923, pp. 197-8.

⁴⁸ Early English Text Society, *The Coventry Leet Book,* ed. M. D. Harris, 1907-13, p. 544.

⁴⁹ Wyclif, *On the Seven Deadly Sins,* chap. xxiv (*Works of Wyclif,* ed. T. Arnold, vol. iii, pp. 154-5). The word rendered "loan" is "leeve" [? leene] in the text.

⁵⁰ For examples of such cases see *Early Chancery Proceedings,* Bdle. lxiv, nos. 291 and 1089; Bdle. xxxvii, no. 38; Bdle. xlvi, no. 307. They are discussed in some detail in my introduction to Thomas Wilson's *Discourse upon Usury,* 1925, pp. 28-9.

⁵¹ *Hist. MSS. Com., MSS. of Marquis of Lothian,* p. 27; Selden Soc., *Leet Jurisdiction in the City of Norwich,* p. 35.

⁵² Aquinas, *Summa Theol.,* 1ᵃ 2ᵃᵉ, Q. xcv, art. ii.

⁵³ *On the Seven Deadly Sins,* chap. xxiv (*Works of Wyclif,* ed. T. Arnold, vol. iii, p. 153) : "Bot men of lawe and marchauntis and chapmen and vitelers synnen more in avarice then done pore laboreres. And this token hereof; for now ben thei pore, and now ben thei ful riche, for wronges that thei done."

⁵⁴ *E.g.,* Ægidius Lessinus, *De Usuris,* cap. ix, pt. i: "Tantum res estimatur juste, quantum ad utilitatem possidentis refertur, et tantum juste valet, quantum sine fraude vendi potest. . . . Omnis translatio facta libera voluntate dominorum juste fit;" Johannes Buridanus, *Quæstiones super decem libros Ethicorum Aristotelis,* v, 23: "Si igitur rem suam sic alienat, ipse secundum suam estimationem non damnificatur, sed lucratur; igitur non injustum patitur." Both writers are discussed by Schreiber (*op. cit.,* pp. 161-71 and 177-91). The theory of Buridanus appears extraordinarily modern; but he is careful to emphasize that prices should be fixed "secundum utilitatem et necessitatem totius communitatis," not "penes necessitatem ementis vel vendentis."

⁵⁵ St. Antonino, *Summa Theologica,* pars ii, tit. i, cap. viii, § 1, and cap. xvi, § iii. An account of St. Antonino's theory of prices is given by Ilgner, *Die volkswirthschaftlichen Anschauungen Antonins von Florenz,* chap. iv; Jarrett, *St. Antonino and Mediæval Economics;* and Schreiber, *op. cit.,* pp. 217-23. Its interest consists in the attempts to maintain the principle of the just price, while making allowance for practical necessities.

⁵⁶ Henry of Langenstein, *Tractatus bipartitus de contractibus emptionis et venditionis,* i, 11, 12 (quoted Schreiber, *op. cit.,* pp. 198-200).

⁵⁷ For these examples see *Cal. of Early Mayor's Court Rolls of the City of London,* ed. A. H. Thomas, pp. 259-60; *Records of the City of Norwich,* ed. W. Hudson and J. C. Tingey, vol. i, 1906, p. 227; *Cal. of Early Mayor's Court Rolls,* p. 132; J. M. Wilson, *The Worcester Liber Albus,* 1920, pp. 199-200, 212-13. The question of the legitimacy of rent-charges and of the profits of partnership has been fully discussed by Max Neumann, *Geschichte des Wuchers in Deutschland* (1865), and by

Ashley, *Economic History.* See also G. O'Brien, *An Essay on Mediæval Economic Teaching* (1920), and G. G. Coulton, *An Episode in Canon Law* (in *History*, July 1921), where the difficult question raised by the Decretal *Naviganti* is discussed.

[58] *Bernardi Papiensis Summa Decretalium* (ed. E. A. D. Laspeyres, 1860) ; lib. v, tit. xv.

[59] *E.g.,* Ægidius Lessinus, *De Usuris,* cap. ix, pt. ii : "Etiam res futuræ per tempora non sunt tantæ estimationis, sicut eædem collectæ in instanti, nec tantam utilitatem inferunt possidentibus, propter quod oportet, quod sint minoris estimationis secundum justitiam."

[60] O'Brien (*op. cit.*) appears, unless I misunderstand him, to take this view.

[61] *Politics,* I, iii, *ad. fin.* 1258[b]. See *Who said "Barren Metal"?* by E. Cannan, W. D. Ross, etc., in *Economica,* June 1922, pp. 105-7.

[62] Innocent IV, *Apparatus,* lib. v, *De Usuris.*

[63] For Italy, see Arturo Segre, *Storia del Commercio,* vol. i, pp. 179-91, and for France, P. Boissonade, *Le Travail dans l'Europe chrétienne au Moyen Age,* 1921, pp. 206-9, 212-13. Both emphasize the financial relations of the Papacy.

[64] *E.g.,* Council of Arles, 314; Nicæa, 325; Laodicea, 372; and many others.

[65] *Corpus Juris Canonici,* Decretal. Greg. IX, lib. v, tit. xix, cap. i.

[66] *Ibid.,* cap. iii.

[67] *Ibid.,* Sexti Decretal, lib. v, tit. v, cap. i, ii.

[68] *Ibid.,* Clementinarum, lib. v, tit. v, cap. i.

[69] The passages referred to in this paragraph are as follows : *Corp. Jur. Can.,* Decretal. Greg. IX, lib. v, tit. xix, cap. ix, iv, x, xiii, xv, ii, v, vi.

[70] *A Formulary of the Papal Penitentiary in the Thirteenth Century,* ed. H. C. Lea, 1892, Nos. xcii, clxxviii (2), clxxix.

[71] *Raimundi de Penna-forti Summa Pastoralis* (Ravaisson, *Catalogue Général des MSS. des Bibliothèques publiques des Departements,* 1849, vol. i, pp. 592 *seqq.*). The archdeacon is to inquire : "Whether [the priest] feeds his flock, assisting those who are in need and above all those who are sick. Works of mercy also are to be suggested by the archdeacon, to be done by him for their assistance. If he cannot fully accomplish them out of his own resources, he ought, according to his power, to use his personal influence to get from others the means of carrying them out. . . . Inquiries concerning the parishioners are to be made, both from the priest and from others among them worthy of credence, who, if necessary, are to be summoned for the purpose to the presence of the archdeacon, as well as from the neighbours, with regard to matters which appear to need correction. First, inquiry is to be made whether there are notorious usurers, or persons reputed to be usurers, and what sort of usury they practise, whether any one, that is to say, lends money or anything else . . . on condition that he receive anything above the principal, or holds any pledge and takes profits from it in excess of the principal, or receives pledges and uses them in the meantime for his own gain; . . . whether he holds horses in pledge and reckons in the cost of their fodder more than they can eat . . . or

whether he buys anything at a much lower price than it is worth, on condition that the seller can take it back at a fixed term on paying the price, though the buyer knows that he (the seller) will not be able to do so; or whether he buys anything for a less price than it is worth, because he pays before receiving the article, for example, standing corn; or whether any one, as a matter of custom and without express contract, is wont to take payment above the principal, as the Cahorsines do. . . . Further, it is to be inquired whether he practises usury cloaked under the guise of a partnership (*nomine societatis palliatam*), as when a man lends money to a merchant, on condition that he be a partner in the gains, but not in the losses. . . . Further, whether he practises usury cloaked under the guise of a penalty, that is to say, when his intention in imposing a penalty [for non-payment at a given date] is not that he may be paid more quickly, but that he may be paid more. Further, whether he practises usury in kind, as when a rich man, who has lent money, will not receive from a poor man any money above the principal, but agrees that he shall work two days in his vineyard, or something of the kind. Further, whether he practises usury cloaked by reference to a third party, as when a man will not lend himself, but has a friend whom he induces to lend. When it has been ascertained how many persons in that parish are notorious for usury of this kind, their names are to be reduced to writing, and the archdeacon is to proceed against them in virtue of his office, causing them to be cited to his court on a day fixed, either before himself or his responsible official, even if there is no accuser, on the ground that they are accused by common report. If they are convicted, either because their offence is evident, or by their own confession, or by witnesses, he is to punish them as he thinks best. . . . If they cannot be directly convicted, by reason of their manifold shifts and stratagems, nevertheless their ill fame as usurers can easily be established. . . . If the archdeacon proceed with caution and diligence against their wicked doings, they will hardly be able to hold their own or to escape—if, that is . . . he vex them with trouble and expense, and humiliate them, by frequently serving citations on them and assigning several different days for their trial, so that by trouble, expense, loss of time, and all manner of confusion they may be induced to repent and submit themselves to the discipline of the Church."

[72] E. Martène and U. Durand, *Thesaurus novus Anecdotorum,* 1717, vol. iv, pp. 696 *seqq.*

[73] Pecock, *The Repressor of over-much blaming of the Clergy,* ed. C. Babington, 1860, pt. i, chap. iii, pp. 15-16. His words show both the difficulties which confronted ecclesiastical teaching and the attempts to overcome them. "I preie thee . . . seie to me where in Holi Scripture is yoven the hundrid parti of the teching upon matrimonie which y teche in a book mad upon *Matrimonie,* and in the firste partie of *Cristen religioun* . . . Seie to me also where in Holi Scripture is yoven the hundrid part of the teching which is yoven upon usure in the thridde parti of the book yclepid *The filling of the iiij tables;* and yit al thilk hool teching yoven upon usure in the now named book is litil ynough or ouer litle for to leerne, knowe and have sufficientli into mannis behove and into Goddis trewe service and lawe keping what is to be leerned and

kunnen aboute usure, as to reeders and studiers ther yn it muste needis be open. Is ther eny more writen of usure in al the Newe Testament save this, Luke vi, 'Geve ye loone, hoping no thing ther of,' and al that is of usure writen in the Oold Testament favourith rather usure than it reproveth. How evere, therfore, schulde eny man seie that the sufficient leernyng and kunnyng of usure or of the vertu contrarie to usure is groundid in Holi Scripture? Howe evere schal thilk litil now rehercid clausul, Luke vi, be sufficient for to answere and assoile alle the harde scrupulose doutis and questiouns which al dai han neede to be assoiled in mennis bargenyngis and cheffaringis togidere? Ech man having to do with suche questiouns mai soone se that Holi Writt geveth litil or noon light thereto at al. Forwhi al that Holi Writt seith ther to is that he forbedith usure, and therfore al that mai be take therbi is this, that usure is unleeful; but though y bileeve herbi that usure is unleeful, how schal y wite herbi what usure is, that y be waar for to not do it, and whanne in a bargeyn is usure, though to summen seemeth noon, and how in a bargeyn is noon usure though to summen ther semeth to be?"

Pecock's defence of the necessity of commentaries on the teaching of Scripture was the real answer to the statement afterwards made by Luther that the text, "Love thy neighbour as thyself," was an all-sufficient guide to action (see Chap. II, p. 99). Examples of teaching as to usury contained in books such as Pecock had in mind will be found in Myrc's *Instructions for Parish Priests* (Early English Text Society, ed. E. Peacock and F. J. Furnivall, 1902), the *Pupilla Oculi,* and *Dan Michel's Ayenbite of Inwyt* (Early English Text Society, ed. R. Morris, 1866).

[74] *The Catechism of John Hamilton, Archbishop of St. Andrews,* 1552, ed. T. G. Law, 1884, pp. 97-9. Under the seventh commandment are denounced: "Fyftlie, al thay that defraudis or spoulyeis the common geir, aganis the common weill for lufe of their awin pryvate and singulare weill. Saxtlie, all usuraris and ockiraris synnis aganis this command, that wil nocht len thair geir frelie, bot makis conditione of ockir, aganis the command of Christe. Sevintlie, all thay quhilk hais servandis or work men and wyll nocht pay theim thair fee or waige, accordyng to conditioun and thair deservyng, quilk syn, as sanct James sayis, cryis vengeance before God. Auchtlie, all thai that strykis cowyne of unlauchful metall, quhair throuch the common weil is hurt and skaithit. The nynte, all Merchandis that sellis corruppit and evyll stufe for gude, and gyf thay or ony uther in bying or sellyng use desait, falsate, parjurie, wrang mettis or weychtis, to the skaith of thair nychtbour, thay committ gret syn agane this command. Nother can we clenge fra breakyng of this command all kyndis of craftis men quhilk usis nocht thair awin craft leillalie and trewlie as thai suld do. . . . All wrechis that wyl be ground ryche incontynent, quhay be fraud, falset, and gyle twynnis men and thair geir, quhay may keip thair nychbour fra povertie and myschance and dois it nocht. Quhay takis ouer sair mail, ouer mekle ferme or ony blake maillis fra thair tennands, or puttis thair cottaris to ouir sair labouris, quhair throw the tenentis and cottaris is put to herschip. Quha invies his nychbouris gud fortune, ouir byis him or takis his geir out of his handis with fair hechtis, or prevenis him,

or begyles him at his marchandis hand." The detail in which different forms of commercial sharp practice are denounced is noticeable.

[75] See *e.g.* Matt. Paris, *Chron. Maj.,* vol. iii, pp. 191-2, for the case of a priest who, for refusing to give Christian burial to an excommunicate usurer, is seized by order of the Count of Brittany and buried alive, bound to the dead man. See also *Materials for the History of Thomas Becket,* vol. v, p. 38.

[76] Harduin, *Acta Conciliorum,* vol. vii, pp. 1017-20; "Anno prædicto [1485], diebus Mercurii et Jovis prædictis, scilicet ante Ramos Palmarum, ibidem apud Vicanum, in claustro ecclesiæ de Vicano; coram domino archiepiscopo, et mandato suo, personæ infrascriptæ, parochiani de Guorgonio, qui super usuraria pravitate erant quam plurimum diffamati; coram domino propter hoc vocati abjuraverunt: et per mandatum domini summas infrascriptas, quas se confessi fuerunt habuisse per usurariam pravitatem, per juramentum suum restituere promiserunt, et stare juri super his coram eo. Bertrandus de Faveriis abjuratus usuras, ut præmittitur, promisit restituere centum solidos monetæ antiquæ: quos, prout ipse confessus est, habuerat per usurariam pravitatem. . . ." Thirty-six more cases were treated in this way.

[77] Villani, *Cronica,* book xii, chap. lviii (ed. 1823, vol. vi, p. 142): Villani complains of the conduct of the inquisitor: "Ma per attignere danari, d'ogni piccola parola oziosa che alcuno dicesse per iniquità contra Iddio, o dicesse che usura non fosse peccato mortale, o simili parole, condannava in grossa somma di danari, secondo che l'uomo era ricco."

[78] Constitutions of Clarendon, cap. 15: "Placita de debitis, quæ fide interposita debentur, vel absque interpositione fidei, sint in justitia regis." On the whole subject see Pollock and Maitland, *History of English Law,* 2nd ed., 1898, vol. ii, pp. 197-202, and F. Makower, *Constitutional History of the Church of England,* 1895, § 60.

[79] *Cal. of Early Mayor's Court Rolls of the City of London,* ed. A. H. Thomas, pp. 44, 88, 156, 235; Selden Soc., *Borough Customs,* ed. M. Bateson, vol. ii, 1906, pp. 161 (London) and 209-10 (Dublin); *Records of Leicester,* ed. M. Bateson, vol. ii, 1901, p. 49. For similar prohibitions by manorial courts, see *Hist. MSS. Com., MSS. of Marquis of Lothian,* p. 28, and G. P. Scrope, *History of the Manor and Barony of Castle Combe,* 1852, p. 238.

[80] *Annales de Burton,* p. 256; Wilkins, *Concilia,* vol. ii, p. 115; *Rot. Parl.,* vol. ii, p. 129*b*.

[81] *Cal. of Letter Books of the City of London,* ed. R. R. Sharpe, vol. H, pp. 23-4, 24-5, 27, 28, 200, 206-7, 261-2, 365; *Liber Albus,* bk. iii, pt. ii, pp. 77, 315, 394-401, 683; Selden Soc., *Leet Jurisdiction in the City of Norwich,* p. 35; *Hist. MSS. Com., MSS. of Marquis of Lothian,* pp. 26, 27.

[82] *Rot. Parl.,* vol. ii, pp. 332*a*, 350*b*.

[83] R. H. Morris, *Chester in the Plantagenet and Tudor Reigns,* 1894 (?), p. 190.

[84] Early Chancery Proceedings, Bdle. xi, no. 307; Bdle. xxix, nos. 193-5; Bdle. xxxi, nos. 96-100, 527; Bdle. lx, no. 20; Bdle. lxiv, no. 1089. See also *Year Books and Plea Rolls as Sources of Historical In-*

*formation*, by H. G. Richardson, in *Trans. Royal Historical Society*, 4th series, vol. v, 1922, pp. 47-8.

⁸⁵ Ed. Gibson, *Codex Juris Ecclesiastici Anglicani*, 2nd ed., 1761, p. 1026.

⁸⁶ 15 Ed. III, st. 1, c. 5; 3 Hen. VII, c. 5; 11 Hen. VII, c. 8; 13 Eliz. c. 8; 21 Jac. I, c. 17.

⁸⁷ *Cal. of Early Mayor's Court Rolls of City of London*, ed. A. H. Thomas, pp. 1, 12, 28-9, 33-4, 44, 52, 88, 141, 156, 226, 235, 251. The cases of the smiths and spurriers occur on pp. 33-4 and 52. In the fifteenth century a gild still occasionally tried to enforce its rules by proceedings in an ecclesiastical court (see Wm. H. Hale, *A Series of Precedents and Proceedings in Criminal Causes*, 1847, nos. xxxvi and lxviii, where persons breaking gild rules are cited before the Commissary's court).

⁸⁸ Canterbury and York Soc., *Registrum Thome Spofford*, ed. A. T. Bannister, 1919, p. 52 (1424); and Surtees Society, vol. cxxxviii, *The Register of Thomas of Corbridge, Lord Archbishop of York*, ed. Wm. Brown, 1925, vol. i, pp. 187-8: "6 kal. Maii, 1303. Wilton.' Littera testimonialis super purgacione domini Johannis de Multhorp, vicarii ecclesie de Garton', de usura sibi imposita. Universis Christi fidelibus, ad quos presentes littere pervenerint, pateat per easdem quod, cum dominus Johannes de Multhorp', vicarius ecclesie de Garton', nostre diocesis, coram nobis Thoma, Dei gracia, etc., in visitacione nostra super usura fuisset notatus, videlicet, quod mutuavit cuidam Jollano de Briddale, ut dicebatur, xxxiij s. iiij d., eo pacto quod idem vicarius ab eo reciperet per x annos annis singulis x s. pro eisdem, de quibus eciam dictum fuit quod prefatus Jollanus dicto vicario pro octo annis ex pacto satisfecit et solvit predicto; eundem vicarium super hoc vocari fecimus coram nobis et ei objecimus supradicta, que ipse inficians constancius atque negans se optulit in forma juris super hiis legitime purgaturum. Nos autem eidem vicario purgacionem suam cum sua sexta manu vicariorum et aliorum presbiterorum sui ordinis indiximus faciendam, quam die Veneris proxima ante festum apostolorum Philippi et Jacobi (April 26), anno gracie m°ccc° tercio, ad hoc sibi prefixo, in manerio nostro de Wilton' super articulo recipimus supradicto, idemque vicarius unacum dominis Johanne, rectore ecclesie B.M. juxta portam castri de Eboraco, Johanne et Johanne, de Wharrum et de Wyverthorp' ecclesiarum vicariis ac Roberto, Johanne, Alano, Stepheno et Willelmo, de Nafferton', Driffeld', Wetewang', Foston' et Wintringham ecclesiarum presbiteris parochialibus fidedignis, de memorato articulo legitime se purgavit; propter quod ipsum vicarium sic purgatum pronunciamus et inmunem sentencialiter declaramus, restituentes eundem ad suam pristinam bonam famam. In cujus rei testimonium sigillum nostrum presentibus est appensum."

⁸⁹ *Early Chancery Proceedings*, Bdle. xviii, no. 137; Bdle. xix, no. 2155; Bdle. xxiv, no. 255; Bdle. xxxi, no. 348. See also A. Abram, *Social England in the Fifteenth Century*, 1909, pp. 215-17. In view of these examples, it seems probable that a more thorough examination of the *Early Chancery Proceedings* would show that, even in the fifteenth century, the jurisdiction of the ecclesiastical courts in matters of contract and usury was of greater practical importance than has sometimes been supposed.

[90] Surtees Soc., vol. lxiv, 1875 (*Acts of Chapter of the Collegiate Church of Ripon*) contains more than 100 cases in which the court deals with questions of contract, debt, etc. The case which is dismissed "propter civilitatem causæ" occurs in 1532 (Surtees Soc., vol. xxi, 1845, *Ecclesiastical Proceedings from the Courts of Durham*, p. 49).

[91] Chetham Soc., vol. xliv, 1901, *Act Book of the Ecclesiastical Court of Whalley*, pp. 15-16.

[92] Surtees Soc., vol. lxiv, 1875, *Acts of Chapter of the Collegiate Church of Ripon*, p. 26.

[93] Hale, *op. cit.* (note 87 above), no. ccxxxviii.

[94] See Chap. III, p. 161.

[95] For parishes, see S. O. Addy, *Church and Manor*, 1913, chap. xv, where numerous examples are given. For a gild which appears to have acted as a bank, see *Hist. MSS. Com.*, 11th Report, 1887, Appx., pt. iii, p. 228 (*MSS. of the Borough of King's Lynn*), and for other examples of loans, H. F. Westlake, *The Parish Gilds of Mediæval England*, 1919, pp. 61-3, *Records of the City of Oxford*, ed. Wm. H. Turner, 1880, p. 8, *Statutes of Lincoln Cathedral*, ed. C. Wordsworth, pt. ii, 1897, pp. 616-17, and G. Unwin, *The Gilds and Companies of London*, 1908, p. 121. For a hospital, see *Hist. MSS. Com.*, 14th Report, Appx., pt. viii, 1895, p. 129 (*MSS. of the Corporation of Bury St. Edmunds*), where 20d. is lent (or given) to a poor man to buy seed for his land. A statement (made half a century after the Dissolution) as to loans by monasteries is quoted by F. A. Gasquet, *Henry VIII and the English Monasteries*, 7th ed., 1920, p. 463; specific examples are not known to me.

[96] W. H. Bliss, *Cal. of Papal Letters*, vol. i, pp. 267-8.

[97] For the early history of the *Monts de Piété* see Holzapfel, *Die Anfänge der Montes Pietatis* (1903), and for their development in the Low Countries, A. Henne, *Histoire du Règne de Charles-quint en Belgique*, 1859, vol. v, pp. 220-3. For proposals to establish them in England see S.P.D. Eliz., vol. cx, no. 57 (printed in Tawney and Power, *Tudor Economic Documents*, vol. iii, sect. iii, no. 6), and my introduction to Thomas Wilson's *Discourse upon Usury*, 1925, pp. 125-7.

[98] Camden Soc., *A Relation of the Island of England about the Year 1500* (translated from the Italian), 1847, p. 23.

[99] Lyndwood, *Provinciale*, sub. tit. *Usura*, and Gibson, *Codex Jur. Eccl. Angl.*, vol. ii, p. 1026.

[100] Pecock, *The Repressor of over-much blaming of the Clergy*, pt. iii, chap. iv, pp. 296-7: "Also Crist seide here in this present proces, that 'at God' it is possible a riche man to entre into the kingdom of heuen; that is to seie, with grace which God profrith and geueth . . . though he abide stille riche, and though withoute such grace it is ouer hard to him being riche to entre. Wherfore folewith herof openli, that it is not forbodun of God eny man to be riche; for thanne noon such man schulde euere entre heuen. . . . And if it be not forbode eny man to be riche, certis thanne it is leeful ynough ech man to be riche; in lasse than he vowe the contrarie or that he knowith bi assay and experience him silf so miche indisposid anentis richessis, that he schal not mowe rewle him silf aright anentis tho richessis: for in thilk caas he is bonde to holde him silf in poverte." The embarrassing qualification at the end—which

suggests the question, who then dare be rich?—is the more striking because of the common-sense rationalism of the rest of the passage.

[101] Trithemius, quoted by J. Janssen, *History of the German People at the close of the Middle Ages,* vol. ii, 1896, p. 102.

[102] *Cal. of Early Mayor's Court Rolls of the City of London,* ed. A. H. Thomas, pp. 157-8.

[103] See A. Luchaire, *Social France at the time of Philip Augustus* (translated by E. B. Krehbiel), pp. 391-2, where an eloquent denunciation by Jacques de Vitry is quoted.

[104] *Topographer and Genealogist,* vol. i, 1846, p. 35. (The writer is a surveyor, one Humberstone.)

[105] See e.g. Chaucer, *The Persone's Tale,* §§ 64-6. The parson expresses the orthodox view that "the condicioun of thraldom and the firste cause of thraldom is for sinne." But he insists that serfs and lords are spiritually equal: "Thilke that thou clepest thy thralles been goddes peple; for humble folk been Cristes freendes."

[106] Gratian, *Decretum,* pt. ii, causa x, Q. ii, c. iii, and causa xii, Q. ii, c. xxxix.

[107] *Summa Theol.,* I$^a$ 2$^{ae}$, Q. xciv, art. v, § 3.

[108] An article of the German Peasants' program in 1525 declared: "For men to hold us as their own property . . . is pitiable enough, considering that Christ has delivered and redeemed us all, the lowly as well as the great, without exception, by the shedding of His precious blood. Accordingly it is consistent with Scripture that we should be free." (The program is printed in J. S. Schapiro, *Social Reform and the Reformation,* 1909, pp. 137-42.) The rebels under Ket prayed "that all bondmen may be made free, for God freed them all with His precious blood-shedding" (printed in Bland, Brown, and Tawney, *English Economic History, Select Documents,* pt. ii, sect. i, no. 8).

# CHAPTER II

[1] *A Lecture on the Study of History,* delivered at Cambridge, June 11, 1895, by Lord Acton, p. 9.

[2] W. Sombart (*Der moderne Kapitalismus,* 1916, vol. i, pp. 524-6) gives facts and figures. See also J. Strieder, *Studien zur Geschichte kapitalistischer Organisationsformen,* 1914, kap. i, ii.

[3] E. R. Daenell, *Die Blütezeit der Deutschen Hanse,* 1905; Schanz, *Englische Handelspolitik gegen die Ende des Mittelalters,* vol. i; N. S. B. Gras, *The Early English Customs System,* 1918, pp. 452-514.

[4] *E.g., The Fugger News-Letters, 1568–1605,* ed. V. von Klarwill, trans. P. de Chary, 1924.

[5] E. Albèri, *Le Relazione degli Ambasciatori Veneti al Senato,* serie I, vol. iii, 1853, p. 357 (*Relazione di Filippo II Re di Spagna da Michele Soriano nel 1559*) : "Questi sono li tesori del re di Spagna, queste le miniere, queste l'Indie che hanno sostentato l'imprese dell' Imperatore tanti anni."

[6] The best contemporary picture of the trade of Antwerp is that of L. Guicciardini, *Descrittione di tutti i Paesi Bassi* (1567), of which part is reprinted in a French translation in Tawney and Power, *Tudor Economic Documents,* vol. iii, pp. 149-173. The best modern accounts of Antwerp are given by Pirenne, *Histoire de Belgique,* vol. ii, pp. 399-403, and vol. iii, pp. 259-72; Ehrenberg, *Das Zeitalter der Fugger,* vol. ii, pp. 3-68; and J. A. Goris, *Étude sur les Colonies Marchandes Méridionales à Anvers de 1488 à 1567* (1925).

[7] The Meutings had opened a branch in Antwerp in 1479, the Hoch-stetters in 1486, the Fuggers in 1508, the Welsers in 1509 (Pirenne, *op. cit.,* vol. iii, p. 261).

[8] Pirenne, *op. cit.,* vol. iii, pp. 273-6.

[9] Ehrenberg, *op. cit.,* vol. ii, pp. 7-8.

[10] A short account of international financial relations in the sixteenth century will be found in my introduction to Thomas Wilson's *Discourse upon Usury,* 1925, pp. 60-86.

[11] Erasmus, *Adagia;* see also *The Complaint of Peace.*

[12] For the Fuggers, see Ehrenberg, *op. cit.,* vol. i, pp. 85-186, and for the other German firms mentioned, *ibid.,* pp. 187-269.

[13] See Goris, *op. cit.,* pp. 510-45, where the reply of the Paris theologians is printed in full; and Ehrenberg, *op. cit.,* vol. ii, pp. 18, 21. For Bellarmin, see Goris, *op. cit.,* pp. 551-2. A curious illustration of the manner in which it was still thought necessary in the later sixteenth century, and in Protestant England, to reconcile economic policy with canonist doctrine, will be found in S.P.D. Eliz., vol. lxxv, no. 54 (printed in Tawney and Power, *Tudor Economic Documents,* vol. iii, pp. 359-70). The writer, who is urging the repeal of the Act of 1552

forbidding all interest whatever, cites Aquinas and Hostiensis to prove
that "trewe and unfayned interest" is not to be condemned as usury.

[14] Ashley, *Economic History*, 1893, vol. i, pt. ii, pp. 442-3.

[15] Bodin, *La Response de Jean Bodin aux Paradoxes de Malestroit
touchant l'enchérissement de toutes choses et le moyen d'y remédier.*

[16] See Max Neumann, *Geschichte des Wuchers in Deutschland*, 1865,
pp. 487 *seqq.*

[17] Calvin's views will be found in his *Epistolæ et Responsa*, 1575, pp.
355-7, and in Sermon xxviii in the Opera.

[18] Bucer, *De Regno Christi.*

[19] *Third Decade*, 1st and 2nd Sermons, in *The Decades of Henry Bullinger* (Parker Society), vol. iii, 1850.

[20] Luther, *Kleiner Sermon vom Wucher* (1519) in *Werke* (Weimar
ed.), vol. vi, pp. 1-8; *Grosser Sermon vom Wucher* (1520), in *ibid.*,
pp. 33-60; *Von Kaufshandlung und Wucher* (1524), in *ibid.*, vol. xv, pp.
279-322; *An die Pfarrherrn wider den Wucher zu predigen, Vermahnung*
(1540), in *ibid.*, vol. li, pp. 325-424.

[21] "*Hie müsste man wahrlich auch den Fuckern und der geistlichen
Gesellschaft einen Zaum ins Maul legen*" (quoted by Ehrenberg, *op. cit.*,
vol. i, p. 117 n.).

[22] See pp. 114-15.

[23] Luther, *Wider die räuberischen und mörderischen Rotten der
Bauern* (1525), in *Werke*, vol. xviii, pp. 357-61.

[24] Latimer, *Sermons;* Ponet, *An Exhortation, or rather a Warning, to
the Lords and Commons;* Crowley, *The Way to Wealth,* and *Epigrams*
(in *Select Works of Robert Crowley*, ed. J. M. Cowper, E.E.T.S., 1872);
Lever, *Sermons*, 1550 (English Reprints, ed. E. Arber, 1895); Becon,
*The Jewel of Joy*, 1553; Sandys, 2nd, 10th, 11th, and 12th of *Sermons*
(Parker Society, 1841); Jewel, *Works*, pt. iv, pp. 1293-8 (Parker Society, 1850). Citations from less well-known writers and preachers will
be found in J. O. W. Haweis, *Sketches of the Reformation*, 1844.

[25] Gairdner, *Letters and Papers of Henry VIII*, vol. xvi, no. 357.

[26] Bossuet, *Traité de l'Usure.* For an account of his views, see Favre,
*Le prêt à intérêt dans l'ancienne France.*

[27] *Brief Survey of the Growth of Usury in England with the Mischiefs attending it*, 1673.

[28] For an account of these changes see K. Lamprecht, *Zum Verständ-
niss der wirthschaftlichen und sozialen Wandlungen in Deutschland vom
14. zum 16. Jahrhundert*, in the *Zeitschrift für Sozial- und Wirthschaftsgeschichte*, Bd. i, 1893, pp. 191 *seqq.*

[29] Lamprecht, *op. cit.*, and J. S. Schapiro, *Social Reform and the Reformation*, 1909, pp. 40-73.

[30] Schapiro, *op. cit.*, pp. 20-39, and Strieder, *op. cit.* (see note 2), pp.
156-212.

[31] For the so-called Reformation of the Emperor Sigismund see Chap.
I, note 24, and for the Peasants' Articles, *ibid.*, note 108.

[32] For Geiler von Kaiserberg and Hipler see Schapiro, *op. cit.*, pp.
30, 126-31. For Hutten see H. Wiskemann, *Dartstellung der in Deutschland zur Zeit der Reformation herrschenden Nationalökonomischen
Ansichten*, 1861, pp. 13-24.

[33] Quoted W. Raleigh, *The English Voyages of the Sixteenth Century*, 1910, p. 28.

[34] Troeltsch, *Protestantism and Progress*, 1912, pp. 44-52.

[35] Schapiro, *op. cit.*, p. 137.

[36] See citations in Wiskemann, *op. cit.*, pp. 47-8, and, for a discussion of Luther's social theory, Troeltsch, *Die Soziallehren der Christlichen Kirchen*, 1912, pp. 549-93.

[37] Luther, *An den christlichen Adel deutscher Nation* (1520), in *Werke*, vol. vi, pp. 381 *seqq.*

[38] Schapiro, *op. cit.*, p. 139.

[39] Luther, *Ermahnung zum Frieden auf die zwölf Artikel der Bauernschaft in Schwaben* (1525), in *Werke*, vol. xviii, p. 327.

[40] *Von Kaufshandlung und Wucher*, in *ibid.*, vol. xv, p. 295.

[41] *An den christlichen Adel*, in *ibid.*, vol. vi, p. 466 (quoted by R. H. Murray, *Erasmus and Luther*, 1920, p. 239).

[42] *Von Kaufshandlung und Wucher*, in *ibid.*, vol. xv, pp. 293-4, 312.

[43] *Concerning Christian Liberty*, in Wace and Buchheim, *Luther's Primary Works*, 1896, pp. 256-7.

[44] *Grosser Sermon vom Wucher*, in *Werke*, vol. vi, p. 49.

[45] See note 73 on Chapter I.

[46] Printed in Neumann, *Geschichte des Wuchers in Deutschland*, Beilage F, pp. 618-19.

[47] *Concerning Christian Liberty*, in Wace and Buchheim, *op. cit.*, pp. 258-9.

[48] *Von Kaufshandlung und Wucher*, in *Werke*, vol. xv, p. 302.

[49] Zwingli, *Von der göttlichen und menschlichen Gerechtigkeit, oder von dem göttlichen Gesetze und den bürgerlichen Gesetzen*, printed in R. Christoffel, *H. Zwingli, Leben und ausgewählte Schriften*, 1857, pt. ii, pp. 313 *seqq.* See also Wiskemann, *op. cit.*, pp. 71-4.

[50] "Quid si igitur ex negociatione plus lucri percipi possit quam ex fundi cuiusvis proventu? Unde vero mercatoris lucrum? Ex ipsius inquies, diligentia et industria" (quoted by Troeltsch, *Die Soziallehren der Christlichen Kirche*, p. 707).

[51] Bucer, *De Regno Christi*.

[52] Roger Fenton, *A Treatise of Usurie*, 1612, p. 61.

[53] Calvin, *Institutes of the Christian Religion*, trans. by J. Allen, 1838, vol. ii, p. 147 (bk. iii, ch. xxiii, par. 7).

[54] *Ibid.*, vol. ii, pp. 128-9 (bk. iii, ch. xxi, par. 7).

[55] Gerrard Winstanley, *A New-Yeer's Gift for the Parliament and Armie*, 1650 (Thomason Tracts, Brit. Mus., E. 587 [6], p. 42).

[56] *The Works of William Laud, D.D.*, ed. Wm. Scott, vol. vi, pt. i, 1857, p. 213.

[57] *De Subventione Pauperum*.

[58] "Quod ad maiores natu spectat, a nobis quotannis repetitur inspectio cuiusque familiæ. Distribuimus inter nos urbis regiones, ut ordine singulas decurias executere liceat. Adest ministro comes unus ex senioribus. Illic novi incolæ examinantur. Qui semel recepti sunt, omittuntur; nisi quod requiritur sitne domus pacata et recte composita, num lites cum vicinis, num qua ebrietas, num pigri sint et ignari ad conciones frequentendas" (quoted by Wiskemann, *op. cit.*, p. 80 n.). For his condemnation of indiscriminate alms-giving, see *ibid.*, p. 79 n.

[59] *De non habendo Pauperum Delectu* (1523), and *De Erogatione Eleemosynarum* (1524). See K. R. Hagenbach, *Johann Oekolampad und Oswald Myconius, die Reformatoren Basels*, 1859, p. 46.

[60] Carl Pestallozzi, *Heinrich Bullinger, Leben und ausgewählte Schriften*, 1858, pp. 50-1, 122-5, 340-2.

[61] Wiskemann, *op. cit.*, pp. 70-4.

[62] Quoted by Preserved Smith, *The Age of the Reformation*, 1921, p. 174.

[63] Calvin, *Inst.*, bk. iv, ch. xii, par. I.

[64] Printed in Paul Henry, *Das Leben Johann Calvins*, vol. ii, 1838, *Appx.*, pp. 26-41.

[65] R. Christoffel, *Zwingli, or the Rise of the Reformation in Switzerland*, trans. by John Cochran, 1858, pp. 159-60.

[66] Printed in Paul Henry, *op. cit.*, vol. ii, Appx., pp. 23-5.

[67] E. Choisy, *L'Etat Chrétien Calviniste à Genève au temps de Théodore de Bèze*, 1902, p. 145. I should like to make acknowledgments to this excellent book for most of the matter contained in the following paragraphs.

[68] Paul Henry, *op. cit.*, pp. 70-5. Other examples are given by Preserved Smith, *op. cit.*, pp. 170-4, and by F. W. Kampschulte, *Johann Calvin, seine Kirche und sein Staat in Genf*, 1869. Statistical estimates of the bloodthirstiness of Calvin's régime vary; Smith (p. 171) states that in Geneva, a town of 16,000 inhabitants, 58 persons were executed and 76 banished in the years 1542-6.

[69] Knox, quoted by Preserved Smith, *op. cit.*, p. 174.

[70] Calvin, *Inst.*, bk. iii, ch. vii, par. 5.

[71] Choisy, *op. cit.*, pp. 442-3.

[72] *Ibid.*, pp. 35-37.

[73] *Ibid.*, pp. 189, 117-19.

[74] *Ibid.*, pp. 35, 165-7.

[75] *Ibid.*, pp. 119-21.

[76] *Ibid.*, pp. 189-94.

[77] Paul Henry, *op. cit.*, vol. ii, p. 70 n.

[78] See the description of the Church given in Calvin, *Inst.*, bk. iv, ch. i, par. 4: "Quia nunc de ecclesia visibili disserere propositum est, discamus vel matris elogio, quam utilis sit nobis eius cognitio, immo necessaria, quando non alius est in vitam ingressus nisi nos ipsa concipiat in utero, nisi pariat, nisi nos alat suis uberibus, denique sub custodia et gubernatione sua nos tueatur, donec excuti carne mortali, similes erimus angelis. Neque enim patitur nostra infirmitas a schola nos dimitti, donec toto vitæ cursu discipuli fuerimus. Adde quod extra eius gremium nulla est speranda peccatorum remissio nec ulla salus."

[79] John Quick, *Synodicon in Gallia Reformata: Or the Acts, Decisions, Decrees and Canons of those famous National Councils of the Reformed Churches in France*, 1692, vol. i, p. 99.

[80] *Ibid.*, vol. i, p. 9 (pirates and fraudulent tradesmen), pp. 25, 34, 38, 79, 140, 149 (interest and usury), p. 70 (false merchandize and selling of stretched cloth), p. 99 (reasonable profits), pp. 162, 204 (investment of money for the benefit of the poor), pp. 194, 213 (lotteries).

81 *The Buke of Discipline*, in *Works of John Knox*, ed. D. Laing, vol. ii, 1848, p. 227.

82 Scottish History Soc., *St. Andrews Kirk Session Register*, ed. D. H. Fleming, 1889-90, vol. i, p. 309; vol. ii, p. 822.

83 W. B. Weeden, *Economic and Social History of New England*, 1890, vol. i, p. 11. The words are Governor Bradford's.

84 *Winthrop's Journal "History of New England," 1630-49*, ed. J. K. Hosmer, 1908, vol. i, pp. 134, 325; vol. ii, p. 20.

85 Weeden, *op. cit.*, vol. i, pp. 125, 58.

86 Winthrop, *op. cit.*, vol. ii, p. 20.

87 J. A. Doyle, *The English in America*, vol. ii, 1887, p. 57; the price of cattle "must not be judged by urgent necessity, but by reasonable profit."

88 Roger Williams, *The Bloudy Tenent of Persecution*, 1644, chap. lv.

89 Winthrop, *op. cit.*, vol. i, pp. 315-18. A similar set of rules as to the conduct of the Christian in trade are given by Bunyan in *The Life and Death of Mr. Badman*, 1905 ed., pp. 118-22.

90 I owe this phrase to the excellent book of J. T. Adams, *The Founding of New England*.

# CHAPTER III

[1] J. Rossus, *Historia Regum Angliæ* (ed. T. Hearne).

[2] 4 Hen. VII, c. 19; 6 Hen. VIII, c. 5; 7 Hen. VIII, c. 1; 25 Hen. VIII, c. 13. For the Commission of 1517 see Leadam, *The Domesday of Enclosures.*

[3] For examples see J. S. Schapiro, *Social Reform and the Reformation,* pp. 60-1, 65, 67, 70-1.

[4] More, *Utopia,* p. 32 (Pitt Press ed., 1879): "Noblemen and gentlemen, yea and certeyne abbottes, holy men no doubt . . . leave no grounde for tillage, thei enclose al into pastures." For a case of claiming a bondman see Selden Society, vol. xvi, 1903, *Select Cases in the Court of Star Chamber,* pp. cxxiii-cxxix, 118-29 (Carter *v.* the Abbott of Malmesbury); for conversion of copyholds to tenancies at will, Selden Society, vol. xii, 1898, *Select Cases in the Court of Requests,* pp. lix-lxv, 64-101 (Kent and other inhabitants of Abbot's Ripton *v.* St. John; the change was alleged to have been made in 1471).

[5] A. Savine, *English Monasteries on the Eve of the Dissolution* (*Oxford Studies in Social and Legal History,* ed. P. Vinogradoff, vol. i, 1909, p. 100), estimates the net temporal income of English monasteries in 1535 at £109,736, and the net income from all sources at £136,361. These figures require to be multiplied by at least 12 to convert them into terms of modern money. An estimate of the capital value which they represent can only be a guess, but it can hardly have been less (in terms of modern money) than £20,000,000.

[6] For the status and payments of grantees, see the figures of Savine, printed in H. A. L. Fisher, *The Political History of England, 1485-1547,* Appx. ii: the low price paid by peers is particularly striking. The best study is that of S. B. Liljegren, *The Fall of the Monasteries and the Social Changes in England leading up to the Great Revolution* (1924), which shows in detail (pp. 118-25) the activities of speculators.

[7] *Star Chamber Proc.,* Hen. VIII, vol. vi, no. 181, printed in Tawney and Power, *Tudor Economic Documents,* vol. i, pp. 19-29.

[8] Selden Society, *Select Cases in the Court of Requests,* pp. lviii-lxix, 198-200.

[9] Quoted by F. A. Gasquet, *Henry the Eighth and the English Monasteries,* 1920, pp. 227-8.

[10] See, *e.g., The Obedience of a Christian Man* (in Tyndale's *Doctrinal Treatises,* Parker Society, 1848), p. 231, where the treatment of the poor by the early Church is cited as an example; and *Policies to reduce this Realme of Englande unto a Prosperus Wealthe and Estate,* 1549 (printed in Tawney and Power, *Tudor Economic Documents,* vol. iii, pp. 311-45): "Like as we suffered our selfes to be ignorant of the trewe worshipping of God, even so God kepte from us the right knowledge how to reforme those inconveniences which we did see before our eyes to tende unto the utter Desolation of the Realme. But now that

the trew worshepping of Gode is . . . so purely and sincerely sett forthe, it is likewise to be trusted that God . . . will use the kinges maiestie and your grace to be also his ministres in plucking up by the roots all the cawses and occasions of this foresaid Decaye and Desolation."

[11] Bucer, *De Regno Christi.*

[12] A. F. Leach, *The Schools of Mediæval England,* 1915, p. 331. He goes on: "The contrasts between one grammar school to every 5,625 people, and that presented by the Schools Inquiry Report in 1864 of one to every 23,750 people . . . is not to the disadvantage of our pre-Reformation ancestors." For details of the Edwardian spoliation, see the same author's *English Schools at the Reformation, 1546-8* (1896).

[13] See *Acts of the Privy Council,* vol. ii, pp. 193-5 (1548) ; in response to protests from the members for Lynn and Coventry, the gild lands of those cities are regranted to them.

[14] Crowley, *The Way to Wealth,* in *Select Works of Robert Crowley,* ed. J. M. Cowper (Early English Text Society, 1872, pp. 129-150).

[15] Crowley, *op. cit.,* and *Epigrams* (in *ibid.,* pp. 1-51).

[16] Becon, *The Jewel of Joy,* 1553: "They abhore the names of Monkes, Friers, Chanons, Nonnes, etc., but their goodes they gredely gripe. And yet where the cloysters kept hospitality, let out their fermes at a resonable price, norished scholes, brought up youth in good letters, they do none of all these thynges."

[17] Thomas Lever, *Sermons,* 1550 (*English Reprints,* ed. E. Arber, 1895), p. 32. The same charge is repeated in subsequent sermons.

[18] F. W. Russell, *Kett's Rebellion in Norfolk,* 1859, p. 202. For Somerset's policy and the revolt of the gentry against it, see Tawney, *The Agrarian Problem in the Sixteenth Century,* pp. 365-70.

[19] Latimer, *Seven Sermons before Edward VI* (*English Reprints,* ed. E. Arber, 1895), pp. 84-6.

[20] *Pleasure and Pain,* in *Select Works of Robert Crowley,* ed. J. M. Cowper, p. 116.

[21] *The Way to Wealth,* in *ibid.,* p. 132.

[22] Lever, *op. cit.,* p. 130.

[23] *A Prayer for Landlords,* from *A Book of Private Prayer set forth by Order of King Edward VI.*

[24] Bacon, *Of the True Greatness of the Kingdom of Britain.*

[25] For a discussion of the problem of credit as it affected the peasant and small master, see my introduction to Wilson's *Discourse upon Usury,* 1925, pp. 17-30.

[26] See note 71 on Chapter I.

[27] D'Ewes, *Journals,* 1682, p. 173.

[28] *Calendar S.P.D. Eliz.,* vol. cclxxxvi, nos. 19, 20.

[29] For examples see S. O. Addy, *Church and Manor,* 1913, chap. xv. The best account of parish business and organization is given by S. L. Ware, *The Elizabethan Parish in its Ecclesiastical and Financial Aspects,* 1908.

[30] Lever, *op. cit.,* p. 130. See also Harrison, *The Description of Britaine,* 1587 ed., bk. ii, chap. xviii.

[31] *A Godlie Treatise concerning the Lawful Use of Riches,* a translation by Thos. Rogers from the Latin of Nicholas Heming, 1578, p. 8.

[32] Sandys, 2nd, 10th, 11th, and 12th of *Sermons* (Parker Society, 1841); Jewel, *Works,* pt. iv, pp. 1293-8 (Parker Society, 1850); Thos. Wilson, *A Discourse upon Usury,* 1572; Miles Mosse, *The Arraignment and Conviction of Usurie,* 1595; John Blaxton, *The English Usurer, or Usury Condemned by the Most Learned and Famous Divines of the Church of England,* 1634.

[33] Heming, *op. cit.,* pp. 16-17.

[34] Roger Fenton, *A Treatise of Usurie,* 1612, p. 59.

[35] Wilson, *op. cit.,* 1925 ed., p. 281.

[36] Miles Mosse, *op. cit.*

[37] *S.P.D. Eliz.,* vol. lxxv, no. 54. (Printed in Tawney and Power, *Tudor Economic Documents,* vol. iii, pp. 359-70).

[38] Heming, *op. cit.,* p. 11.

[39] Maitland, *English Law and the Renaissance,* 1901.

[40] Quoted by Maitland, *op. cit.,* pp. 49-50.

[41] Wilson, *op. cit.*

[42] Jeremy Taylor, *Ductor Dubitantium,* 1660, bk. iii, ch. iii, par. 30.

[43] Mosse, *op. cit.,* Dedication, p. 6.

[44] E. Cardwell, *Synodalia,* 1842, p. 436.

[45] Cardwell, *The Reformation of the Ecclesiastical Laws,* 1850, pp. 206, 323.

[46] *The Remains of Archbishop Grindal,* ed. Wm. Nicholson (Parker Soc., 1843), p. 143.

[47] See, *e.g.,* W. P. M. Kennedy, *Elizabethan Episcopal Administration,* 1924, vol. iii, p. 180 (Archdeacon Mullins' Articles for the Archdeaconry of London (1585): "Item, whether you do know that within your parish there is (or are) any person or persons notoriously known or suspected by probable tokens or common fame to be an usurer; or doth offend by any colour or means directly or indirectly in the same"), and pp. 184, 233; Wilkins, *Concilia,* vol. iv, pp. 319, 337, 416.

[48] Cardwell, *Synodalia,* vol. i, pp. 144, 308; Wilkins, *Concilia,* vol. iv, p. 509.

[49] Ware, *op. cit.* (see note 29 above), quotes several examples. See also *Archæologia Cantiana,* vol. xxv, 1902, pp. 27, 48 (Visitations of the Archdeacon of Canterbury).

[50] *Hist. MSS. Com., 13th Report,* 1892, Appx., pt. iv, pp. 333-4 (*MSS. of the Borough of Hereford*).

[51] W. H. Hale, *A Series of Precedents and Proceedings in Criminal Causes,* 1847, p. 166.

[52] *Yorkshire Arch. Journal,* vol. xviii, 1895, p. 331.

[53] *Commissary of London Correction Books, 1618-1625* (H. 184, pp. 164, 192). I am indebted to Mr. Fincham of Somerset House (where the books are kept) for kindly calling my attention to these cases. The shorter of them (p. 192) runs as follows:

| Sancti Botolphi extra Aldersgate Thomas Witham at the signe of the Unicorne | Detected for an usurer that taketh above the rate of $x^{li}$ in the $100^{li}$ and above the rate of 2s. in the pound for money by him lent for a yeare, or more than after that rate for a lesse tyme ex fama prout in rotula. Quo die comparuit, etc. |

9mo Maii 1620 coram domino officiali principali etc. et in eius camera etc. comparuit dictus Witham et ei objecto ut supra allegavit that he is seldom at home himselfe but leaves his man to deale in the business of his shop, and yf any fault be committed he saith the fault is in his man and not in himselfe, and he sayeth he will give charge and take care that no oppression shall be made nor offence committed this way hereafter, humbly praying the judge for favour to be dismissed, unde dominus monuit eum that thereafter neither by himselfe nor his servant he offende in the lyke nor suffer any such oppression to be committed, et cum hac monitione eum dimisit.

[54] *S.P.D. Eliz.,* vol. lxxv, no. 54.

[55] For an account of these expedients see my introduction to Wilson's *Discourse upon Usury,* 1925, pp. 123-8.

[56] Richard Hooker, *The Laws of Ecclesiastical Policy,* bk. viii, chap. i, par. 5.

[57] *Acts of the Privy Council,* vol. xxvii, 1597, p. 129.

[58] *The Stiffkey Papers* (ed. H. W. Saunders, Royal Historical Society, Camden Third Series, vol. xxvi, 1915), p. 140.

[59] Quoted by E. M. Leonard, *The Early History of English Poor Relief,* 1900, p. 148.

[60] For an account of the treatment of exchange business under Elizabeth, see Wilson, *op. cit.,* Introduction, pp. 146-54. ·

[61] For references see *ibid.,* pp. 164-5; and *Les Reportes del Cases in Camera Stellata, 1593-1609,* ed. W. P. Baildon, 1894, pp. 235-7. The latter book contains several instances of intervention by the Star Chamber in cases of engrossing of corn (pp. 71, 76-7, 78-9, 91) and of enclosure and depopulation (pp. 49-52, 164-5, 192-3, 247, 346-7).

[62] *A Discourse of the Common Weal of this Realm of England,* ed. E. Lamond, 1893, p. 14.

[63] *The Works of William Laud, D.D.,* ed. Wm. Scott, vol. i, 1847, p. 6.

[64] *Ibid.,* p. 64.

[65] *Ibid.,* pp. 89, 138.

[66] *Ibid.,* p. 167.

[67] *Ibid.,* pp. 28-9.

[68] Gonner, *Common Land and Enclosure,* 1912, pp. 166-7. For the activity of the Government from 1629 to 1640, see Tawney, *The Agrarian Problem in the Sixteenth Century,* pp. 376, 391, and E. M. Leonard, *The Inclosure of Common Fields in the Seventeenth Century,* in *Trans. Royal Hist. Soc.,* N.S., vol. xix, pp. 101 *seqq.*

[69] Letter to Dr. Gilbert Sheldon, Warden of All Souls (in Laud's *Works,* vol. vi, pt. ii, p. 520): "One thing more I must tell you, that, though I did you this favour, to make stay of the hearing till your return, yet for the business itself, I can show you none; partly because I am a great hater of depopulations in any kind, as being one of the greatest mischiefs in this kingdom, and of very ill example from a college, or college tenant"; Clarendon, *History of the Rebellion,* bk. i, par. 204.

[70] *S.P.D. Chas. I,* vol. ccccxcix, no. 10 (printed in Tawney, *The Agrarian Problem in the Sixteenth Century,* pp. 420-1); and *Lords' Journals,* vol. vi, p. 468b (March 13, 1643-4), Articles against Laud: "Then Mr. Talbot upon oath deposed how the Archbishop did oppose

the law in the business of inclosures and depopulations; how, when the law was desired to be pleaded for the right of land, he bid them 'Go plead law in inferior Courts, they should not plead it before him'; and that the Archbishop did fine him for that business two hundred pounds for using the property of his freehold, and would not suffer the law to be pleaded."

[71] Leonard, *The Early History of English Poor Relief*, pp. 150-64; Unwin, *Industrial Organization in the Sixteenth and Seventeenth Centuries*, 1904, pp. 142-7.

[72] R. R. Reid, *The King's Council in the North*, 1921, pp. 412, 413 n.

[73] Camden Soc., N.S., vol. xxxix, 1886, *Cases in the Courts of Star Chamber and High Commission*, ed. S. R. Gardiner, p. 46. For another case of engrossing of corn, see *ibid.*, pp. 82-9.

[74] Tawney, *The Assessment of Wages in England by the Justices of the Peace*, in *Vierteljahrschrift für Sozial- und Wirthschaftsgeschichte*, Bd. xi, 1913, pp. 551-4; Leonard, *op. cit.*, p. 157.

[75] *The Works of William Laud*, ed. Wm. Scott, vol. vi, 1857, pt. i, p. 191. (Answer to Lord Saye and Sele's speech upon the Bill about Bishops' Powers in Civil Affairs and Courts of Judicature.)

[76] *Ibid.*, vol. i, pp. 5-6.

[77] Harrington, *Works*, 1700 ed., pp. 69 (*Oceana*) and 388-9 (*The Art of Law-giving*).

[78] G. Malynes, *Lex Mercatoria*, 1622. The same simile had been used much earlier in *A Discourse of the Common Weal of this Realm of England*, ed. E. Lamond, p. 98.

[79] D'Ewes, *Journals*, p. 674; and 39 Eliz., c. 2.

[80] For criticisms of price control see Tawney and Power, *Tudor Economic Documents*, vol. iii, pp. 339-41, and vol. ii, p. 188, and *Stiffkey Papers* (see note 58 above), pp. 130-40.

[81] H. Ellis, *Original Letters*, 2nd series, vol. ii, 1827, letter clxxxii, and J. W. Burgon, *The Life and Times of Sir Thomas Gresham*, 1839, vol. ii, p. 343.

[82] Wilson, *op. cit.* (see note 55 above), p. 249.

[83] *Commons' Journals*, May 21, 1604, vol. i, p. 218.

[84] 13 Eliz., c. 8, repealing 5 and 6 Ed. VI, c. 20; D'Ewes, *Journals*, pp. 171-4.

[85] Owen and Blakeway, *History of Shrewsbury*, 1825, vol. ii, pp. 364 n., 412.

[86] *Hist. MSS. Com., Report on MSS. in various Collections*, vol. i, 1901, p. 46 (*MSS. of Corporation of Burford*).

[87] Wilson, *op. cit.* (see note 55 above), p. 233.

[88] Coke, *Institutes,* pt. ii, 1797, pp. 601 *seqq.* (*Certain articles of abuses which are desired to be reformed in granting of prohibitions, exhibited by Richard Bancroft, Archbishop of Canterbury.*)

[89] Thomas Ridley, *A View of the Civile and Ecclesiastical Law, and wherein the Practice of them is streitened and may be relieved within this Land*, 1607, Dedication, p. 3.

[90] W. Huntley, *A Breviate of the Prelates' intolerable Usurpation*, 1637, pp. 183-4. The case referred to is that of Hinde, alleged to have been heard Mich. 18 and 19 Eliz. For the controversy over prohibitions, see

R. G. Usher, *The Rise and Fall of the High Commission*, 1913, pp. 180 *seqq*.

[91] D'Ewes, *Journals*, pp. 171, 173.

[92] See, *e.g.*, Surtees Society, vol. xxxiv, 1858, *The Acts of the High Commission Court within the Diocese of Durham*, Preface, which shows that between 1626 and 1639 cases of contempt of the ordinary ecclesiastical jurisdiction ran into hundreds.

[93] Penn, *No Cross, No Crown*, pt. i, ch. xii, par. 8.

[94] Sanderson, *De Obligatione Conscientiæ*, 1666; Taylor, *The Rule and Exercises of Holy Living*, 1650, chap. iii, sect. iii (*Of Negotiation or Civil Contracts, Rules and Measures of Justice in Bargaining*).

[95] Mandeville, *The Fable of the Bees*, ed. F. B. Kaye, 1924, pp. 193, 194. Similar sentiments with regard to the necessity of poverty were expressed later by the Rev. J. Townsend, in his *Dissertation on the Poor Laws* (1785), and by Patrick Colquhoun in his *Treatise on the Wealth and Resources of the British Empire* (1814). Like Mandeville, both these writers argue that poverty is essential to the prosperity, and, indeed, to the very existence, of civilization. For a full collection of citations to the same effect from eighteenth-century writers, see E. S. Furniss, *The Position of the Laborer in a System of Nationalism*, 1920, chaps. iv-vi.

[96] *The Whole Duty of Man, laid down in a plain and familiar Way for the Use of All*, 1658.

# CHAPTER IV

[1] Tucker, *A Brief Essay on the Advantages and Disadvantages which respectively attend France and Great Britain with regard to Trade*, 1750, p. 33. The best account of Tucker, most of whose works are scarce, is given by W. E. Clark, *Josiah Tucker, Economist* (*Studies in History, Economics and Public Law*, Columbia University, vol. xix, 1903-5).

[2] *Reliquiæ Baxterianæ: or Mr. Richard Baxter's Narrative of the most memorable Passages of his Life and Times*, 1696, p. 5.

[3] Bunyan, *The Pilgrim's Progress*.

[4] *The Life of the Duke of Newcastle*, by Margaret, Duchess of Newcastle (Everyman ed., 1915, p. 153).

[5] Baxter, *op. cit.*, p. 31.

[6] Bunyan, *Pilgrim's Progress*.

[7] Baxter, *op. cit.*, p. 89.

[8] Thomas Fuller, *The Holy and Profane States*, 1884 ed., p. 122.

[9] Quoted S. Seyer, *Memoirs of Bristol*, vol. ii, 1823, p. 314.

[10] R. G. Usher, *The Reconstruction of the English Church*, vol. i, 1910, pp. 249-50.

[11] Baxter, *op. cit.*, p. 30.

[12] *An orderly and plaine Narration of the Beginnings and Causes of this Warre*, 1644, p. 4 (Brit. Mus., Thomason Tracts, E. 54 [3]). I owe this reference to the kindness of Father Paschal Larkin.

[13] Clarendon, *History of the Rebellion*, bk. vi, par. 271.

[14] Parker, *Discourse of Ecclesiastical Politie*, 1670, Preface, p. xxxix.

[15] *The Life of Edward, Earl of Clarendon, written by himself*, 1827 ed., vol. iii, p. 101.

[16] D. C. A. Agnew, *Protestant Exiles from France*, 1886, vol. i, pp. 20-1. In 1640 the Root and Branch Petition included, among the evils due to the Bishops, "the discouragement and destruction of all good subjects, of whom are multitudes, both clothiers, merchants and others, who, being deprived of their ministers, and overburthened with these pressures, have departed the kingdom to Holland and other parts, and have drawn with them a great manufacture of cloth and trading out of the land into other places where they reside, whereby wool, the great staple of the kingdom, is become of small value, and vends not, trading is decayed, many poor people want work, seamen lose employment, and the whole land is much impoverished" (S. R. Gardiner, *Constitutional Documents of the Puritan Revolution, 1628-60* [1889], p. 73). For instances of the comparatively liberal treatment of alien immigrants under Elizabeth, see Tawney and Power, *Tudor Economic Documents*, vol. i, section vi, nos. 3, 4, 11 (2), 15, and Cunningham, *Growth of English Industry and Commerce, Modern Times*, 1921, pt. i, pp. 79-84.

[17] *Toryism and Trade can never agree*, 1713, p. 12. The tract is

wrongly ascribed to Davenant by H. Levy, *Economic Liberalism*, 1913, p. 12.

[18] See, *e.g.*, G. Martin, *La Grande Industrie sous le règne de Louis XIV*, 1899, chap. xvii, where the reports of several intendants are quoted; and Levasseur, *Histoire du commerce de la France*, 1911, vol. i, p. 421.

[19] *A Letter from a Gentleman in the City to a Gentleman in the Country about the Odiousness of Persecution*, 1677, p. 29.

[20] Sir Wm. Temple, *Observations upon the United Provinces of the Netherlands*, chap. v, vi.

[21] *The True Interest and Political Maxims of the Republick of Holland and West-Friesland*, 1702, pt. i, chap. xiv.

[22] Petty, *Political Arithmetic*, 1690, pp. 25-6.

[23] *The Present Interest of England stated, by a Lover of his King and Country*, 1671. I am indebted to Mr. A. P. Wadsworth for calling my attention to the passage quoted in the text. The same point is put more specifically by Lawrence Braddon: "The superstition of their religion obligeth France to keep (at least) fifty Holy days more than we are obliged to keep; and every such day wherein no work is done is one hundred and twenty thousand pounds loss to the deluded people" (*Abstract of the Draft of a Bill for relieving, reforming and employing the Poor*, 1717). See also Defoe, in his *Enquiry into Occasional Conformity*, 1702, pp. 18-19: "We wonder, gentlemen, you will accept our money on your deficient funds, our stocks to help carry on your wars, our loans and credits to your victualling office and navy office. If you would go on to distinguish us, get a law made we shall buy no lands, that we may not be freeholders; and see if you could find money to buy us out. Transplant us into towns and bodies, and let us trade by our selves; let us card, spin, knit, weave and work with and for one another, and see how you'll maintain your own poor without us. Let us fraight our ships apart, keep our money out of your Bank, accept none of our bills, and separate your selves as absolutely from us in civil matters, as we do from you in religious, and see how you can go on without us."

[24] Swift, *Examiner*.

[25] Bolingbroke, *Letter to Sir Wm. Windham*, 1753, p. 21.

[26] *Reliquiæ Baxterianæ* (see note 2), p. 94. He goes on: "The generality of the Master Workmen [i.e., employers] lived but a little better than their journeymen (from hand to mouth), but only that they laboured not altogether so hard."

[27] Voltaire, *Lettres Philosophiques*, no. x, and Montesquieu, *Esprit des Lois*, xix, 27, and xx, 22. See also the remarks to the same effect in D'Argenson, *Considérations sur le Gouvernement de la France*, 1765.

[28] *Brief Survey of the Growth of Usury in England*, 1673.

[29] Marston, *Eastward Ho!*, act i, sc. i.

[30] Clarendon, *History of the Rebellion*, bk. i, par. 163.

[31] Petty, *Political Arithmetic*, 1690, p. 23.

[32] Max Weber, *Die protestantische Ethik und der Geist des Kapitalismus*, first published in the *Archiv für Sozialwissenschaft und Sozialpolitik Statistik*, vols. xx, xxi, and since reprinted in vol. i of his

*Gesammelte Aufsätze zur Religionssoziologie,* 1920; Troeltsch, *Die Soziallehren der Christlichen Kirchen* and *Protestantism and Progress,* 1912; Schulze-Gaevernitz, *Britischer Imperialismus und Englischer Freihandel,* 1906; Cunningham, *Christianity and Economic Science,* 1914, chap. v.

Weber's essay gave rise to much discussion in Germany. Its main thesis—that Calvinism, and in particular English Puritanism, from which nearly all his illustrations are drawn, played a part of preponderant importance in creating moral and political conditions favorable to the growth of capitalist enterprise—appears to be accepted by Troeltsch, *Die Soziallehren der Christlichen Kirchen,* pp. 704 *seqq.* It is submitted to a critical analysis by Brentano (*Die Anfänge des modernen Kapitalismus,* 1916, pp. 117-57), who dissents from many of Weber's conclusions. Weber's essay is certainly one of the most fruitful examinations of the relations between religion and social theory which has appeared, and I desire to acknowledge my indebtedness to it, in particular with reference to its discussion of the economic application given by some Puritan writers to the idea expressed by the word "calling." At the same time, there are several points on which Weber's arguments appear to me to be one-sided and overstrained, and on which Brentano's criticisms of it seem to me to be sound.

Thus (i), as was perhaps inevitable in an essay dealing with economic and social thought, as distinct from changes in economic and social organization, Weber seems to me to explain by reference to moral and intellectual influences developments which have their principal explanation in another region altogether. There was plenty of the "capitalist spirit" in fifteenth-century Venice and Florence, or in south Germany and Flanders, for the simple reason that these areas were the greatest commercial and financial centers of the age, though all were, at least nominally, Catholic. The development of capitalism in Holland and England in the sixteenth and seventeenth centuries was due, not to the fact that they were Protestant powers, but to large economic movements, in particular the Discoveries and the results which flowed from them. Of course material and psychological changes went together, and of course the second reacted on the first. But it seems a little artificial to talk as though capitalist enterprise could not appear till religious changes had produced a capitalist spirit. It would be equally true, and equally one-sided, to say that the religious changes were purely the result of economic movements.

(ii) Weber ignores, or at least touches too lightly on, intellectual movements, which were favorable to the growth of business enterprise and to an individualist attitude towards economic relations, but which had little to do with religion. The political thought of the Renaissance was one; as Brentano points out, Machiavelli was at least as powerful a solvent of traditional ethical restraints as Calvin. The speculations of business men and economists on money, prices and the foreign exchanges were a second. Both contributed to the temper of single-minded concentration on pecuniary gain, which Weber understands by the capitalist spirit.

(iii) He appears greatly to over-simplify Calvinism itself. In the first place, he apparently ascribes to the English Puritans of the seventeenth century the conception of social ethics held by Calvin and his immediate followers. In the second place, he speaks as though all English Puritans in the seventeenth century held much the same view of social duties and expediency. Both suggestions are misleading. On the one hand, the Calvinists of the sixteenth century (including English Puritans) were believers in a rigorous discipline, and the individualism ascribed not unjustly to the Puritan movement in its later phases would have horrified them. The really significant question is that of the causes of the change from the one standpoint to the other, a question which Weber appears to ignore. On the other hand, there were within seventeenth-century Puritanism a variety of elements, which held widely different views as to social policy. As Cromwell discovered, there was no formula which would gather Puritan aristocrats and Levellers, landowners and Diggers, merchants and artisans, buff-coat and his general, into the fold of a single social theory. The issue between divergent doctrines was fought out within the Puritan movement itself. Some won; others lost.

Both "the capitalist spirit" and "Protestant ethics," therefore, were a good deal more complex than Weber seems to imply. What is true and valuable in his essay is his insistence that the commercial classes in seventeenth-century England were the standard-bearers of a particular conception of social expediency, which was markedly different from that of the more conservative elements in society—the peasants, the craftsmen, and many landed gentry—and that that conception found expression in religion, in politics, and, not least, in social and economic conduct and policy.

[33] Cunningham, *The Moral Witness of the Church on the Investment of Money and the Use of Wealth*, 1909, p. 25.

[34] Knox, *The Buke of Discipline*, in *Works*, ed. D. Laing, vol. ii, 1848, pp. 183 *seqq.*; Thos. Cartwright, *A Directory of Church Government* (printed in D. Neal, *History of the Puritans*, 1822, vol. v, Appx. iv); W. Travers, *A Full and Plain Declaration of Ecclesiastical Discipline*, 1574; J. Udall, *A Demonstration of the Trueth of that Discipline which Christe hath prescribed in his Worde for the Government of his Church*, 1589; Bancroft, *Dangerous Positions and Proceedings published and practised within this Iland of Brytaine under Pretence of Reformation and for the Presbyteriall Discipline*, 1593 (part reprinted in R. G. Usher, *The Presbyterian Movement in the Reign of Queen Elizabeth, as illustrated by the Minute Book of the Dedham Classis*, 1905).

[35] Cartwright, *op. cit.*

[36] Usher, *op. cit.*, p. 1.

[37] *Ibid.*, pp. 14-15, for Bancroft's account of the procedure.

[38] Quoted from Baillie's *Letters* by W. A. Shaw, *A History of the English Church during the Civil Wars and under the Commonwealth*, 1900, vol. i, p. 128.

[39] Shaw, *op. cit.*, vol. ii, chap. iii (*The Presbyterian System, 1646-60*). For the practical working of Presbyterian discipline, see Chetham Society, vols. xx, xxii, xxiv, *Minutes of the Manchester Classis*, and vols. xxxvi, xli, *Minutes of the Bury Classis*.

318 NOTES

⁴⁰ See Chap. III, p. 142.

⁴¹ *Puritan Manifestoes*, p. 120, quoted by H. G. Wood, *The Influence of the Reformation on Ideas concerning Wealth and Property*, in *Property, its Rights and Duties*, 1913, p. 142. Mr. Wood's essay contains an excellent discussion of the whole subject, and I should like here to acknowledge my obligations to it. For the views of Knewstub, Smith, and Baro, see the quotations from them printed by Haweis, *Sketches of the Reformation*, 1844, pp. 237-40, 243-6. It should be noted that Baro, while condemning those who, "sitting idle at home, make merchandise only of their money, by giving it out in this sort to needy persons . . . without having any regard of his commodity to whome they give it, but only of their own gain," nevertheless admitted that interest was not always to be condemned. See also Thos. Fuller, *History of the University of Cambridge*, ed. M. Prickett and T. Wright, 1840, pp. 275-6, 288-9, and Cunningham, *Growth of English Industry and Commerce, Modern Times*, 1921 ed., pt. i, pp. 157-8.

⁴² New Shakespeare Society, Series vi, no. 6, 1877-9, Phillip Stubbes's *Anatomy of the Abuses in England*, ed. F. J. Furnivall, pp. 115-16.

⁴³ W. Ames, *De Conscientia et eius iure vel casibus libri quinque*, bk. v, chaps. xliii, xliv. Ames (1576-1633) was educated at Christ's College, Cambridge, tried to settle at Colchester, but was forbidden to preach by the Bishop of London, went to Leyden about 1610, was appointed to the theological chair at Franeker in 1622, where he remained for ten years, and died at Rotterdam.

⁴⁴ *E.g.*, Stubbes, *op. cit.*; Richard Capel, *Temptations, their Nature, Danger, Cure*, 1633; John Moore, *The Crying Sin of England of not caring for the Poor; wherein Inclosure, viz. such as doth unpeople Townes, and uncorn Fields, is arraigned, convicted and condemned*, 1653.

⁴⁵ J. O. Halliwell, *The Autobiography and Correspondence of Sir Simonds D'Ewes*, 1845, vol. i, pp. 206-10, 322, 354; vol. ii, pp. 96, 153-4.

⁴⁶ Usher, *op. cit.* (see note 34 above), pp. 32, 53, 70, 99-100.

⁴⁷ Sept. 26, 1645, it is resolved "that it shall be in the power of the eldership to suspend from the sacrament of the Lord's supper any person that shall be legally attainted of Barratry, Forgery, Extortion, Perjury, or Bribery" (*Commons' Journals*, vol. iv, p. 290).

⁴⁸ Chetham Society, *Minutes of the Bury Presbyterian Classis, 1647-57*, pt. i, pp. 32-3. The Cambridge *classis* (*ibid.*, pt. ii, pp. 196-7) decided in 1657 that the ordinance of Parliament of August 29, 1648 should be taken as the rule of the *classis* in the matter of scandal. The various scandals mentioned in the ordinance included extortion, and the *classis* decided that "no person lawfully convict of any of the foresaid scandalls bee admitted to the Lord's supper without signification of sincere repentance," but it appears (p. 198) to have been mainly interested in witches, wizards, and fortune-tellers.

⁴⁹ *Hist. MSS. Comm., Report on MSS. in various Collections*, vol. i, 1901, p. 132.

⁵⁰ Quoted by F. J. Powicke, *A Life of the Reverend Richard Baxter*, 1924, p. 92.

⁵¹ Selections from those parts of *The Christian Directory* which bear on social ethics are printed by Jeannette Tawney, *Chapters from Richard*

*Baxter's Christian Directory,* 1925, in which most of the passages quoted in the text will be found.

[52] *Reliquiæ Baxterianæ* (see note 2), p. 1.

[53] *Life and Death of Mr. Badman* (Cambridge English Classics, 1905), pp. 116-25, where Bunyan discusses at length the ethics of prices.

[54] Carlyle, *Cromwell's Letters and Speeches,* Letter ii.

[55] See on these points Weber, *op. cit.* (note 32 above), p. 94, whose main conclusions I paraphrase.

[56] Milton, *A Defence of the People of England* (1692 ed.), p. xvii.

[57] See, *e.g.,* Thos. Wilson, *A Discourse upon Usury,* Preface, 1925 ed., p. 178: "There bee two sortes of men that are alwayes to bee looked upon very narrowly, the one is the dissemblinge gospeller, and the other is the wilfull and indurate papiste. The first under colour of religion overthroweth all religion, and bearing good men in hande that he loveth playnesse, useth covertelie all deceypte that maye bee, and for pryvate gayne undoeth the common welfare of man. And touching thys sinne of usurie, none doe more openly offende in thys behalfe than do these counterfeite professours of thys pure religion."

[58] Fenton, *A Treatise of Usurie,* 1612, pp. 60-1.

[59] *Brief Survey of the Growth of Usury in England,* 1673.

[60] S. Richardson, *The Cause of the Poor Pleaded,* 1653, Thomason Tracts, E. 703 (9), p. 14. For other references, see note 72 below. For extortionate prices, see Thomason Tracts, E. 399 (6), *The Worth of a Penny, or a Caution to keep Money,* 1647. I am indebted for this and subsequent references to the Thomason Tracts to Miss P. James.

[61] Hooker, Preface to *The Laws of Ecclesiastical Polity,* Everyman ed., 1907, vol. i, p. 128.

[62] Wilson, *op. cit.,* p. 250.

[63] *Memoirs of the Life of Colonel Hutchinson, written by his Widow Lucy,* Everyman ed., 1908, pp. 64-5.

[64] See the references given in note 66.

[65] *The Earl of Strafforde's Letters and Despatches,* by William Knowler, D.D., 1739, vol. ii, p. 138.

[66] No attempt has been made in the text to do more than refer to the points on which the economic interests and outlook of the commercial and propertied classes brought them into collision with the monarchy, and only the most obvious sources of information are mentioned here. For patents and monopolies, including the hated soap monopoly, see G. Unwin, *The Gilds and Companies of London,* 1908, chap. xvii, and W. Hyde Price, *The English Patents of Monopoly,* 1906, chap. xi, and *passim.* For the control of exchange business, *Cambium Regis, or the Office of his Majesties Exchange Royall, declaring and justifying his Majesties Right and the Convenience thereof,* 1628, and Ruding, *Annals of the Coinage,* 1819, vol. iv, pp. 201-10. For the punishment of speculation by the Star Chamber, and for projects of public granaries, Camden Society, N.S., vol. xxxix, 1886, *Reports of Cases in the Courts of Star Chamber and High Commission,* ed. S. R. Gardiner, pp. 43 *seqq.,* 82 *seqq.,* and N. S. B. Gras, *The Evolution of the English Corn Market,* 1915, pp. 246-50. For the control of the textile industry and the reaction against it, H. Heaton, *The Yorkshire Woollen and Worsted Industries,*

1920, chaps. iv, vii; Kate E. Barford, *The West of England Cloth Industry: A seventeenth-century Experiment in State Control,* in the *Wiltshire Archæological and Natural History Magazine,* Dec., 1924, pp. 531-42; R. R. Reid, *The King's Council in the North,* 1921, pt. iv, chap. ii; *Victoria County History, Suffolk,* vol. ii, pp. 263-8. For the intervention of the Privy Council to raise the wages of textile workers and to protect craftsmen, Tawney, *The Assessment of Wages in England by the Justices of the Peace,* in the *Vierteljahrschrift für Sozial- und Wirthschaftsgeschichte,* Bd. xi, 1913, pp. 307-37, 533-64; Leonard, *The Early History of English Poor Relief,* pp. 160-3; *Victoria County History, Suffolk,* vol. ii, pp. 268-9; and Unwin, *Industrial Organization in the Sixteenth and Seventeenth Centuries,* 1904, pp. 142-7. For the Depopulation Commissions, Tawney, *The Agrarian Problem in the Sixteenth Century,* pp. 376, 391. For the squeezing of money from the East India Company and the infringement of its Charter, Shafa'at Ahmad Khan, *The East India Trade in the XVIIth Century,* 1923, pp. 69-73. For the colonial interests of Puritan members, A. P. Newton, *The Colonising Activities of the English Puritans,* 1914, and C. E. Wade, *John Pym,* 1912.

[67] E. Laspeyres, *Geschichte der Volkswirthschaftlichen Anschauungen der Niederländer und ihrer Litteratur zur Zeit der Republik,* 1863, pp. 256-70. An idea of the points at issue can be gathered from the exhaustive (and unreadable) work of Salmasius, *De Modo Usurarum,* 1639.

[68] John Quick, *Synodicon in Gallia Reformata,* 1692, vol. i, p. 99.

[69] For the change of sentiment in America, see Troeltsch, *Protestantism and Progress,* pp. 117-27; for Franklin, *Memoirs of the Life and Writings of Benjamin Franklin,* and Sombart, *The Quintessence of Capitalism,* 1915, pp. 116-21.

[70] Rev. Robert Woodrow (quoted by Sombart, *op. cit.,* p. 149).

[71] John Cooke, *Unum Necessarium or the Poore Man's Case* (1648), which contains a plea for the regulation of prices and the establishment of *Monts de Piété.*

[72] For the scandal caused to the Protestant religion by its alleged condonation of covetousness, see T. Watson, *A Plea for Alms,* 1658 (Thomason Tracts, E. 2125), pp. 21, 33-4: "The Church of Rome layes upon us this aspersion that we are against good workes . . . I am sorry that any who go for honest men should be brought into the indightment; I mean that any professors should be impeached as guilty of this sinne of covetousnesse and unmercifulnesse . . . I tell you these devout misers are the reproach of Christianity . . . I may say of penurious votaries, they have the wings of profession by which they seem to fly to heaven, but the feet of beasts, walking on the earth and even licking the dust . . . Oh, take heed, that, seeing your religion will not destroy your covetousnesse, at last your covetousnesse does not destroy your religion." See also Sir Balthazar Gerbier, *A New Year's Result in favour of the Poore,* 1651 (Thomason Tracts, E. 651 [14]), p. 4: "If the Papists did rely as much on faith as the reformed professors of the Gospel (according to our English tenets) doe, or that the reformed professors did so much practice charity as the Papists doe?"

[73] S. Richardson, *op. cit.* (see note 60 above), pp. 7-8, 10.

[74] The first person to emphasize the way in which the idea of a "calling" was used as an argument for the economic virtues was Weber (see note 32 above), to whose conclusions I am largely indebted for the following paragraphs.

[75] Bunyan, *The Pilgrim's Progress.*

[76] Richard Steele, *The Tradesman's Calling, being a Discourse concerning the Nature, Necessity, Choice, etc., of a Calling in general,* 1684, pp. 1, 4.

[77] *Ibid.,* pp. 21-2.

[78] *Ibid.,* p. 35.

[79] Baxter, *Christian Directory,* 1678 ed., vol. i, p. 336*b*.

[80] Thomas Adams (quoted Weber, *op. cit.,* p. 96 n.).

[81] Matthew Henry, *The Worth of the Soul* (quoted *ibid.,* p. 168 n.).

[82] Baxter, *op. cit.,* vol. i, p. 111*a*.

[83] Steele, *op. cit.,* p. 20.

[84] Baxter, *op. cit.,* vol. i, pp. 378*b*, 108*b*; vol. iv, p. 253*a*.

[85] *Navigation Spiritualized: or a New Compass for Seamen, consisting of xxxii Points:*

$$
of \left\{ \begin{array}{l} \textit{Pleasant Observations} \\ \textit{Profitable Applications and} \\ \textit{Serious Reflections.} \end{array} \right.
$$

*All concluded with so many spiritual poems. Whereunto is now added,*

    i. *A sober conversation of the sin of drunkenness.*
    ii. *The Harlot's face in the scripture-glass, etc.*

*Being an essay towards their much desired Reformation from the horrible and detestable sins of Drunkenness, Swearing, Uncleanness, Forgetfulness of Mercies, Violation of Promises, and Atheistical Contempt of Death.* 1682. The author of this cheerful work was a Devonshire minister, John Flavell, who also wrote *Husbandry Spiritualized, or the Heavenly Use of Earthly Things,* 1669. In him, as in Steele, the Chadband touch is unmistakable. *The Religious Weaver,* apparently by one Fawcett, I have not been able to trace.

[86] Steele, *op. cit.* (see note 76 above).

[87] Bunyan, *The Pilgrim's Progress.*

[88] David Jones, *A Farewell Sermon at St. Mary Woolnoth's,* 1692.

[89] Nicholas Barbon, *A Discourse of Trade,* 1690, ed. by Professor John H. Hollander (*A Reprint of Economic Tracts,* Series ii, no. 1).

[90] The words of a member of the Long Parliament, quoted by C. H. Firth, *Oliver Cromwell,* 1902, p. 313.

[91] *The Life of Edward, Earl of Clarendon,* 1827 ed., vol. ii, p. 235: "The merchants took much delight to enlarge themselves upon this argument [i.e., the advantages of war], and shortly after to discourse 'of the infinite benefit that would accrue from a barefaced war against the Dutch, how easily they might be subdued and the trade carried by the English.'" According to Clarendon, who despised the merchants and

hated the whole business, it was almost a classical example of a commercial war, carefully stage-managed in all its details, from the directorship which the Royal African Company gave to the Duke of York down to the inevitable "incident" with which hostilities began.

[92] *Ibid.*, vol. iii, pp. 7-9.

[93] Sir Dudley North, *Discourses upon Trade,* 1691, Preface.

[94] Petty, *Political Arithmetic,* Preface.

[95] Chamberlayne, *Angliæ Notitia* (quoted P. E. Dove, *Account of Andrew Yarranton,* 1854, p. 82 n.).

[96] Roger North, *The Lives of the Norths* (1826 ed.), vol. iii, p. 103; T. Watson, *A Plea for Alms* (Thomason Tracts, E. 2125), p. 33; Dryden, *Absalom and Achitophel,* 2nd part, 1682, p. 9, where Sir Robert Clayton, Lord Mayor 1679-80, and Member of Parliament for the City 1679-81 and again from 1689, appears as "extorting Ishban." He was a scrivener who had made his money by usury.

[97] John Fawke, Sir William Thompson, William Love, and John Jones.

[98] Charles King (*The British Merchant,* 1721, vol. i, p. 181) gives the following persons as signatories of an analysis of the trade between England and France in 1674: Patience Ward, Thomas Papillon, James Houblon, William Bellamy, Michael Godfrey, George Toriano, John Houblon, John Houghe, John Mervin, Peter Paravicine, John Dubois, Benj. Godfrey, Edm. Harrison, Benj. Delaune. The number of foreign names is remarkable.

[99] For Dutch capital in London, see *Hist. MSS. Comm., 8th Report,* 1881, p. 134 (proceedings of the Committee on the decay of trade, 1669); with regard to investment of foreign capital in England, it was stated that "Alderman Bucknell had above £100,000 in his hands, Mr. Meynell above £30,000, Mr. Vandeput at one time £60,000, Mr. Dericost always near £200,000 of Dutch money, lent to merchants at 7, 6, and 5 per cent."

[100] *The Life of Edward, Earl of Clarendon,* vol. ii, pp. 289-93, and vol. iii, pp. 4-7; and John Beresford, *The Godfather of Downing Street,* 1925.

[101] S. Bannister, *William Paterson, the Merchant-Statesman, and Founder of the Bank of England: His Life and Trials,* 1858.

[102] A. Yarranton, *England's Improvement,* 1677.

[103] *The Complete English Tradesman* (1726) belongs to the same genus as the book of Steele (see above, pp. 244-6), but it has reduced Christianity to even more innocuous proportions: see *Letter xvii (Of Honesty in Dealing).*

[104] T. S. Ashton, *Iron and Steel in the Industrial Revolution,* 1924, pp. 211-26. Mr. A. P. Wadsworth has shown that the leading Lancashire clothiers were often Nonconformists (*History of the Rochdale Woollen Trade,* in *Trans. Rochdale Lit. and Sci. Soc.,* vol. xv, 1925).

[105] Quoted F. J. Powicke, *Life of Baxter,* 1924, p. 158.

[106] Dicey, *Law and Public Opinion in England,* 1905, pp. 400-1.

[107] *The Humble Petition of Thousands of well-affected Persons inhabiting the City of London, Westminster, the Borough of Southwark, Hamlets, and Places adjacent* (Bodleian Pamphlets, The Levellers' Petitions, c. 15, 3 Linc.). See also G. P. Gooch, *English Democratic Ideas in the Seventeenth Century,* 1898.

[108] Camden Society, *The Clarke Papers,* ed. C. H. Firth, 1891-4, vol. ii, pp. 217-21 (letter from Winstanley to Fairfax and the Council of War, Dec. 8, 1649).

[109] *Records of the Borough of Leicester, 1603-88,* ed. Helen Stocks, 1923, pp. 370, 414, 428-30.

[110] John Moore, *op. cit.* (see note 44, above), p. 13. See also E. C. K. Gonner, *Common Land and Enclosure,* 1912, pp. 53-5.

[111] Camden Society, *The Clarke Papers,* vol. i, pp. 299 *seqq.,* lxvii *seqq.*

[112] *The Diary of Thomas Burton,* ed. J. T. Rutt, 1828, vol. i, pp. 175-6. A letter from Whalley, referring to agitations against enclosure in Warwickshire, Nottinghamshire, Lincolnshire and Leicestershire, will be found in Thurloe, *State Papers,* vol. iv, p. 686.

[113] Joseph Lee, *A Vindication of a Regulated Enclosure,* 1656, p. 9.

[114] Aquinas, *Summa Theol.,* 2ª 2ᵃᵉ, Q. xxxii, art. v.

[115] *Dives et Pauper,* 1493, Prol., chap. vii; cf. Pecock, *The Repressor of over-much Blaming of the Clergy,* pt. iii, chap. iv, pp. 296-7. For an excellent account of the medieval attitude towards the poor, see B. L. Manning, *The People's Faith in the Time of Wyclif,* 1919, chap. x.

[116] *A Lyke-wake Dirge,* printed by W. Allingham, *The Ballad Book,* 1907, no. xxxi.

[117] Latimer, *The fifth Sermon on the Lord's Prayer* (in *Sermons,* Everyman ed., p. 336). Cf. Tyndale, *The Parable of the Wicked Mammon* (in *Doctrinal Treatises of William Tyndale,* Parker Society, 1848, p. 97): "If thy brother or neighbour therefore need, and thou have to help him, and yet showest not mercy, but withdrawest thy hands from him, then robbest thou him of his own, and art a thief."

[118] Christopher Harvey, *The Overseer of the Poor* (in G. Gilfillan, *The Poetical Works of George Herbert,* 1853, pp. 241-3).

[119] J. E. B. Mayor, *Two Lives of N. Ferrar, by his brother John and Dr. Jebb,* p. 261 (quoted by B. Kirkman Gray, *A History of English Philanthropy,* 1905, p. 54).

[120] *A True Report of the Great Cost and Charges of the foure Hospitals in the City of London,* 1644 (quoted, *ibid.,* p. 66).

[121] See, *e.g., Hist. MSS. Comm., Reports on MSS. in various Collections,* vol. i, 1901, pp. 109-24; Leonard, *Early History of English Poor Relief,* pp. 268-9.

[122] Sir Matthew Hale, *A Discourse touching Provision for the Poor,* 1683.

[123] *Stanley's Remedy, or the Way how to reform wandering Beggars, Thieves, Highway Robbers and Pick-pockets,* 1646 (Thomason Tracts, E. 317 [6]), p. 4.

[124] *Commons' Journals,* March 19, 1648/9, vol. vi, p. 167.

[125] *Ibid.,* vol. vi, pp. 201, 374, 416, 481; vol. vii, p. 127.

[126] Samuel Hartlib, *London's Charity Inlarged,* 1650, p. i.

[127] Hartlib, *op. cit.*

[128] Firth and Rait, *Acts and Ordinances of the Interregnum,* 1911, vol. ii, pp. 104-10. An ordinance creating a corporation had been passed Dec. 17, 1647 (*ibid.,* vol. i, pp. 1042-5).

[129] *Ibid.,* vol. ii, pp. 1098-9.

[130] Stockwood, at Paul's Cross, 1578 (quoted by Haweis, *Sketches of the Reformation*, p. 277).

[131] Steele, *op. cit.* (note 76 above), p. 22.

[132] R. Younge, *The Poores' Advocate*, 1654 (Thomason Tracts, E. 1452 [3]), p. 6.

[133] For these and other passages from Restoration economists to the same effect, see a striking article by Dr. T. E. Gregory on *The Economics of Employment in England (1660-1713)* in *Economica*, no. i, Jan., 1921, pp. 37 *seqq.*, and E. S. Furniss, *The Position of the Labourer in a System of Nationalism*, 1920, chaps. v, vi.

[134] *Das Kommunistische Manifest*, 1918 ed., pp. 27-8: "Die Bourgeoisie, wo sie zur Herrschaft gekommen, hat alle feudalen, patriarchalischen, idyllischen verhältnisse zerstört. Sie hat die buntscheckigen Feudalbande, die den Menschen an seinen natürlichen Vorgesetzten knüpften, unbarmherzig zerrissen, und kein anderes Band zwischen Mensch und Mensch übrig gelassen, als das nackte Interesse, als die gefühllose bare Zahlung."

[135] Defoe, *Giving Alms no Charity*, 1704, pp. 25-7.

[136] Petty, *Political Arithmetic*, p. 45.

[137] Sir Henry Pollexfen, *Discourse of Trade*, 1697, p. 49; Walter Harris, *Remarks on the Affairs and Trade of England and Ireland*, 1691, pp. 43-4; *The Querist*, 1737 (in *The Works of George Berkeley, D.D.*, ed. A. C. Fraser, 1871, p. 387); Thomas Alcock, *Observations on the Defects of the Poor Laws*, 1752, pp. 45 *seqq.* (quoted Furniss, *op. cit.*, p. 153).

[138] Arthur Young, *Eastern Tour*, 1771, vol. iv, p. 361.

[139] Harrison, *The Description of Britaine*, 1587 ed., bk. ii, chap. x, *Of Provision made for the Poor.*

[140] H. Hunter, *Problems of Poverty: Selections from the . . . Writings of Thomas Chalmers, D.D.*, 1912, p. 202.

[141] For the influence of Chalmers' idea on Senior, and, through him, on the new Poor Law of 1834, see T. Mackay, *History of the English Poor Law*, vol. iii, 1899, pp. 32-4. Chalmers held that *any* Poor Law was in itself objectionable. Senior, who described Chalmers' evidence before the Committee on the State of the Poor in Ireland as "the most instructive, perhaps, that ever was given before a Committee of the House of Commons," appears to have begun by agreeing with him, but later to have adopted the principle of deterrence, backed by the test workhouse, as a second best. The Commissioners of 1832-4 were right in thinking the existing methods of relief administration extremely bad; they were wrong in supposing distress to be due mainly to lax administration, instead of realizing, as was the fact, that lax administration had arisen as an attempt to meet the increase of distress. Their discussion of the causes of pauperism is, therefore, extremely superficial, and requires to be supplemented by the evidence contained in the various contemporary reports (such, *e.g.*, as those on the hand-loom weavers) dealing with the industrial aspects of the problem.

[142] W. C. Braithwaite, *The Second Period of Quakerism*, 1919, pp.

560-2.   Defoe comments on the strict business standards of the Quakers in *Letter xvii* (*Of Honesty in Dealing*) in *The Complete English Tradesman*.   Mr. Ashton (*Iron and Steel in the Industrial Revolution*, p. 219) remarks, "The eighteenth century Friend no less than the medieval Catholic held firmly to some doctrine of Just Price," and quotes examples from the conduct of Quaker iron-masters.

# INDEX

Abbot's Ripton, 140, 308

Acton, Lord, 65

Acts of Parliament:
15 Ed. III, st. 1, c. 5 (1341), 52
37 Hen. VIII, c. 9 (1545), 159
5 and 6 Ed. VI, c. 20 (1552), 159, 180
13 Eliz., c. 8 (1571), 159, 180, 181, 187
39 Eliz., c. 2 (1597), 178

Aegidius Lessinus, 295, 296

Aeneas Silvius, 110

Agriculture, 136-50, 231. See also *Enclosures, Land, Pasture farming, Peasants*

Alcock, Thomas, 270

Alien immigrants, 205, 314

Almsgiving, condemnation of, 111, 114, 265, 305; a duty, 260-1

America, silver of, 68, 74, 135; Calvinism in, 127-32, 227, 238, 320

Ames, 216-7, 318

Amsterdam, 104

Anglicans. See *Clergy* and *Church of England*

Annuities, 42, 217

Antwerp, 72, 73-5, 79, 80, 86, 87, 104, 168, 178, 303; fall of, 77, 176

Apparel, excess in, 115, 127

Aquinas. See *St. Thomas*

Archdeacons, visitations of, 48, 52, 162, 296-7, 310

Aristotle, 44

Asceticism, 17, 18, 19

Ashton, T. S., 252, 325

Aske, 141

Augsburg, 79, 85

Bacon, 148, 151, 185

Baillie, 214

Bancroft, Archbishop, 186, 213, 214

Bank, at Geneva, 120
— of England, 252

Banking, deposit, beginnings of, 176

Barbon, Dr. Nicholas, 247

Barebones, Praise-God, 247

Bargaining, equity in, 152, 159, 181, 183, 188, 221-3, 224, 244, 272. See also *Prices* and *Profits*

Baro, 215, 318

Basle, 120; Council of, 103

Baxter, Richard, 9, 19, 200, 203, 207, 219-24, 226, 242 (quoted), 243 (quoted), 253, 260, 268, 291

Becon, 82, 141, 144

Beggars. See *Almsgiving* and *Vagrancy*

Bellarmin, 80, 303

Bellers, 19, 272

Belloc, H., 92

Bennet, Dr., 153

Benvenuto da Imola, quoted, 11

Berkeley, Bishop, 270, 284

Berne, 120

Berthold, Brother, 225

Beza, 119, 121, 122, 123, 215

Birmingham, 204

Bishops, articles of visitation of, 161; were normally justices, 165; Bill *re* powers of, 174-5, 312; abolition of, 188, 214. See also *Commissary, Court of High Commission* and *Courts, ecclesiastical*

Blaxton, John, cited, 156

Bodin, 81

Böheim, Hans, 81

Bolingbroke, 207

Bologna, University of, 81

Boniface VIII, 19; bull of, 21

Bossuet, 83

Boston, 128-31

Bourges, 50

Braddon, Lawrence, 315

Bradford, 204
—, Governor, 127 (quoted), 128

Brentano, 316

Bristol, 202

Brittany, Count of, 299

Bruges, 73

Bucer, 19, 63, 81, 83, 105, 116, 142, 215

327